USE OF A GRADED APPROACH
IN THE APPLICATION
OF THE SAFETY REQUIREMENTS
FOR RESEARCH REACTORS

The following States are Members of the International Atomic Energy Agency:

AFGHANISTAN
ALBANIA
ALGERIA
ANGOLA
ANTIGUA AND BARBUDA
ARGENTINA
ARMENIA
AUSTRALIA
AUSTRIA
AZERBAIJAN
BAHAMAS
BAHRAIN
BANGLADESH
BARBADOS
BELARUS
BELGIUM
BELIZE
BENIN
BOLIVIA, PLURINATIONAL
  STATE OF
BOSNIA AND HERZEGOVINA
BOTSWANA
BRAZIL
BRUNEI DARUSSALAM
BULGARIA
BURKINA FASO
BURUNDI
CAMBODIA
CAMEROON
CANADA
CENTRAL AFRICAN
  REPUBLIC
CHAD
CHILE
CHINA
COLOMBIA
COMOROS
CONGO
COSTA RICA
CÔTE D'IVOIRE
CROATIA
CUBA
CYPRUS
CZECH REPUBLIC
DEMOCRATIC REPUBLIC
  OF THE CONGO
DENMARK
DJIBOUTI
DOMINICA
DOMINICAN REPUBLIC
ECUADOR
EGYPT
EL SALVADOR
ERITREA
ESTONIA
ESWATINI
ETHIOPIA
FIJI
FINLAND
FRANCE
GABON
GEORGIA

GERMANY
GHANA
GREECE
GRENADA
GUATEMALA
GUYANA
HAITI
HOLY SEE
HONDURAS
HUNGARY
ICELAND
INDIA
INDONESIA
IRAN, ISLAMIC REPUBLIC OF
IRAQ
IRELAND
ISRAEL
ITALY
JAMAICA
JAPAN
JORDAN
KAZAKHSTAN
KENYA
KOREA, REPUBLIC OF
KUWAIT
KYRGYZSTAN
LAO PEOPLE'S DEMOCRATIC
  REPUBLIC
LATVIA
LEBANON
LESOTHO
LIBERIA
LIBYA
LIECHTENSTEIN
LITHUANIA
LUXEMBOURG
MADAGASCAR
MALAWI
MALAYSIA
MALI
MALTA
MARSHALL ISLANDS
MAURITANIA
MAURITIUS
MEXICO
MONACO
MONGOLIA
MONTENEGRO
MOROCCO
MOZAMBIQUE
MYANMAR
NAMIBIA
NEPAL
NETHERLANDS
NEW ZEALAND
NICARAGUA
NIGER
NIGERIA
NORTH MACEDONIA
NORWAY
OMAN
PAKISTAN

PALAU
PANAMA
PAPUA NEW GUINEA
PARAGUAY
PERU
PHILIPPINES
POLAND
PORTUGAL
QATAR
REPUBLIC OF MOLDOVA
ROMANIA
RUSSIAN FEDERATION
RWANDA
SAINT KITTS AND NEVIS
SAINT LUCIA
SAINT VINCENT AND
  THE GRENADINES
SAMOA
SAN MARINO
SAUDI ARABIA
SENEGAL
SERBIA
SEYCHELLES
SIERRA LEONE
SINGAPORE
SLOVAKIA
SLOVENIA
SOUTH AFRICA
SPAIN
SRI LANKA
SUDAN
SWEDEN
SWITZERLAND
SYRIAN ARAB REPUBLIC
TAJIKISTAN
THAILAND
TOGO
TONGA
TRINIDAD AND TOBAGO
TUNISIA
TÜRKİYE
TURKMENISTAN
UGANDA
UKRAINE
UNITED ARAB EMIRATES
UNITED KINGDOM OF
  GREAT BRITAIN AND
  NORTHERN IRELAND
UNITED REPUBLIC
  OF TANZANIA
UNITED STATES OF AMERICA
URUGUAY
UZBEKISTAN
VANUATU
VENEZUELA, BOLIVARIAN
  REPUBLIC OF
VIET NAM
YEMEN
ZAMBIA
ZIMBABWE

The Agency's Statute was approved on 23 October 1956 by the Conference on the Statute of the IAEA held at United Nations Headquarters, New York; it entered into force on 29 July 1957. The Headquarters of the Agency are situated in Vienna. Its principal objective is "to accelerate and enlarge the contribution of atomic energy to peace, health and prosperity throughout the world".

IAEA SAFETY STANDARDS SERIES No. SSG-22 (Rev. 1)

USE OF A GRADED APPROACH IN THE APPLICATION OF THE SAFETY REQUIREMENTS FOR RESEARCH REACTORS

SPECIFIC SAFETY GUIDE

INTERNATIONAL ATOMIC ENERGY AGENCY
VIENNA, 2023

COPYRIGHT NOTICE

All IAEA scientific and technical publications are protected by the terms of the Universal Copyright Convention as adopted in 1952 (Berne) and as revised in 1972 (Paris). The copyright has since been extended by the World Intellectual Property Organization (Geneva) to include electronic and virtual intellectual property. Permission to use whole or parts of texts contained in IAEA publications in printed or electronic form must be obtained and is usually subject to royalty agreements. Proposals for non-commercial reproductions and translations are welcomed and considered on a case-by-case basis. Enquiries should be addressed to the IAEA Publishing Section at:

Marketing and Sales Unit, Publishing Section
International Atomic Energy Agency
Vienna International Centre
PO Box 100
1400 Vienna, Austria
fax: +43 1 26007 22529
tel.: +43 1 2600 22417
email: sales.publications@iaea.org
www.iaea.org/publications

IAEA Library Cataloguing in Publication Data

Names: International Atomic Energy Agency.
Title: Use of a graded approach in the application of the safety requirements for
    research reactors / International Atomic Energy Agency.
Description: Vienna : International Atomic Energy Agency, 2023. | Series: IAEA
    safety standards series, ISSN 1020–525X ; no. SSG-22 (Rev. 1) | Includes
    bibliographical references.
Identifiers: IAEAL 22-01548 | ISBN 978–92–0–142822–6 (paperback : alk. paper) |
    ISBN 978–92–0–142922–3 (pdf) | ISBN 978–92–0–143022–9 (epub)
Subjects: LCSH: Nuclear reactors — Risk assessment. | Nuclear reactors — Safety
    measures. | Radiation — Safety measures.
Classification: UDC 621.039.58 | STI/PUB/2035

FOREWORD

by Rafael Mariano Grossi
Director General

The IAEA's Statute authorizes it to "establish…standards of safety for protection of health and minimization of danger to life and property". These are standards that the IAEA must apply to its own operations, and that States can apply through their national regulations.

The IAEA started its safety standards programme in 1958 and there have been many developments since. As Director General, I am committed to ensuring that the IAEA maintains and improves upon this integrated, comprehensive and consistent set of up to date, user friendly and fit for purpose safety standards of high quality. Their proper application in the use of nuclear science and technology should offer a high level of protection for people and the environment across the world and provide the confidence necessary to allow for the ongoing use of nuclear technology for the benefit of all.

Safety is a national responsibility underpinned by a number of international conventions. The IAEA safety standards form a basis for these legal instruments and serve as a global reference to help parties meet their obligations. While safety standards are not legally binding on Member States, they are widely applied. They have become an indispensable reference point and a common denominator for the vast majority of Member States that have adopted these standards for use in national regulations to enhance safety in nuclear power generation, research reactors and fuel cycle facilities as well as in nuclear applications in medicine, industry, agriculture and research.

The IAEA safety standards are based on the practical experience of its Member States and produced through international consensus. The involvement of the members of the Safety Standards Committees, the Nuclear Security Guidance Committee and the Commission on Safety Standards is particularly important, and I am grateful to all those who contribute their knowledge and expertise to this endeavour.

The IAEA also uses these safety standards when it assists Member States through its review missions and advisory services. This helps Member States in the application of the standards and enables valuable experience and insight to be shared. Feedback from these missions and services, and lessons identified from events and experience in the use and application of the safety standards, are taken into account during their periodic revision.

I believe the IAEA safety standards and their application make an invaluable contribution to ensuring a high level of safety in the use of nuclear technology. I encourage all Member States to promote and apply these standards, and to work with the IAEA to uphold their quality now and in the future.

THE IAEA SAFETY STANDARDS

BACKGROUND

Radioactivity is a natural phenomenon and natural sources of radiation are features of the environment. Radiation and radioactive substances have many beneficial applications, ranging from power generation to uses in medicine, industry and agriculture. The radiation risks to workers and the public and to the environment that may arise from these applications have to be assessed and, if necessary, controlled.

Activities such as the medical uses of radiation, the operation of nuclear installations, the production, transport and use of radioactive material, and the management of radioactive waste must therefore be subject to standards of safety.

Regulating safety is a national responsibility. However, radiation risks may transcend national borders, and international cooperation serves to promote and enhance safety globally by exchanging experience and by improving capabilities to control hazards, to prevent accidents, to respond to emergencies and to mitigate any harmful consequences.

States have an obligation of diligence and duty of care, and are expected to fulfil their national and international undertakings and obligations.

International safety standards provide support for States in meeting their obligations under general principles of international law, such as those relating to environmental protection. International safety standards also promote and assure confidence in safety and facilitate international commerce and trade.

A global nuclear safety regime is in place and is being continuously improved. IAEA safety standards, which support the implementation of binding international instruments and national safety infrastructures, are a cornerstone of this global regime. The IAEA safety standards constitute a useful tool for contracting parties to assess their performance under these international conventions.

THE IAEA SAFETY STANDARDS

The status of the IAEA safety standards derives from the IAEA's Statute, which authorizes the IAEA to establish or adopt, in consultation and, where appropriate, in collaboration with the competent organs of the United Nations and with the specialized agencies concerned, standards of safety for protection of health and minimization of danger to life and property, and to provide for their application.

With a view to ensuring the protection of people and the environment from harmful effects of ionizing radiation, the IAEA safety standards establish fundamental safety principles, requirements and measures to control the radiation exposure of people and the release of radioactive material to the environment, to restrict the likelihood of events that might lead to a loss of control over a nuclear reactor core, nuclear chain reaction, radioactive source or any other source of radiation, and to mitigate the consequences of such events if they were to occur. The standards apply to facilities and activities that give rise to radiation risks, including nuclear installations, the use of radiation and radioactive sources, the transport of radioactive material and the management of radioactive waste.

Safety measures and security measures[1] have in common the aim of protecting human life and health and the environment. Safety measures and security measures must be designed and implemented in an integrated manner so that security measures do not compromise safety and safety measures do not compromise security.

The IAEA safety standards reflect an international consensus on what constitutes a high level of safety for protecting people and the environment from harmful effects of ionizing radiation. They are issued in the IAEA Safety Standards Series, which has three categories (see Fig. 1).

Safety Fundamentals

Safety Fundamentals present the fundamental safety objective and principles of protection and safety, and provide the basis for the safety requirements.

Safety Requirements

An integrated and consistent set of Safety Requirements establishes the requirements that must be met to ensure the protection of people and the environment, both now and in the future. The requirements are governed by the objective and principles of the Safety Fundamentals. If the requirements are not met, measures must be taken to reach or restore the required level of safety. The format and style of the requirements facilitate their use for the establishment, in a harmonized manner, of a national regulatory framework. Requirements, including numbered 'overarching' requirements, are expressed as 'shall' statements. Many requirements are not addressed to a specific party, the implication being that the appropriate parties are responsible for fulfilling them.

Safety Guides

Safety Guides provide recommendations and guidance on how to comply with the safety requirements, indicating an international consensus that it

---

[1] See also publications issued in the IAEA Nuclear Security Series.

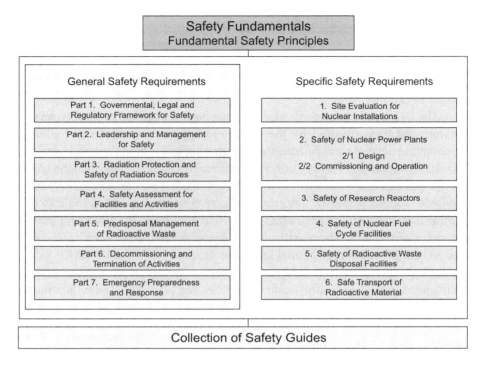

FIG. 1. The long term structure of the IAEA Safety Standards Series.

is necessary to take the measures recommended (or equivalent alternative measures). The Safety Guides present international good practices, and increasingly they reflect best practices, to help users striving to achieve high levels of safety. The recommendations provided in Safety Guides are expressed as 'should' statements.

APPLICATION OF THE IAEA SAFETY STANDARDS

The principal users of safety standards in IAEA Member States are regulatory bodies and other relevant national authorities. The IAEA safety standards are also used by co-sponsoring organizations and by many organizations that design, construct and operate nuclear facilities, as well as organizations involved in the use of radiation and radioactive sources.

The IAEA safety standards are applicable, as relevant, throughout the entire lifetime of all facilities and activities — existing and new — utilized for peaceful purposes and to protective actions to reduce existing radiation risks. They can be used by States as a reference for their national regulations in respect of facilities and activities.

The IAEA's Statute makes the safety standards binding on the IAEA in relation to its own operations and also on States in relation to IAEA assisted operations.

The IAEA safety standards also form the basis for the IAEA's safety review services, and they are used by the IAEA in support of competence building, including the development of educational curricula and training courses.

International conventions contain requirements similar to those in the IAEA safety standards and make them binding on contracting parties. The IAEA safety standards, supplemented by international conventions, industry standards and detailed national requirements, establish a consistent basis for protecting people and the environment. There will also be some special aspects of safety that need to be assessed at the national level. For example, many of the IAEA safety standards, in particular those addressing aspects of safety in planning or design, are intended to apply primarily to new facilities and activities. The requirements established in the IAEA safety standards might not be fully met at some existing facilities that were built to earlier standards. The way in which IAEA safety standards are to be applied to such facilities is a decision for individual States.

The scientific considerations underlying the IAEA safety standards provide an objective basis for decisions concerning safety; however, decision makers must also make informed judgements and must determine how best to balance the benefits of an action or an activity against the associated radiation risks and any other detrimental impacts to which it gives rise.

DEVELOPMENT PROCESS FOR THE IAEA SAFETY STANDARDS

The preparation and review of the safety standards involves the IAEA Secretariat and five Safety Standards Committees, for emergency preparedness and response (EPReSC) (as of 2016), nuclear safety (NUSSC), radiation safety (RASSC), the safety of radioactive waste (WASSC) and the safe transport of radioactive material (TRANSSC), and a Commission on Safety Standards (CSS) which oversees the IAEA safety standards programme (see Fig. 2).

All IAEA Member States may nominate experts for the Safety Standards Committees and may provide comments on draft standards. The membership of the Commission on Safety Standards is appointed by the Director General and includes senior governmental officials having responsibility for establishing national standards.

A management system has been established for the processes of planning, developing, reviewing, revising and establishing the IAEA safety standards.

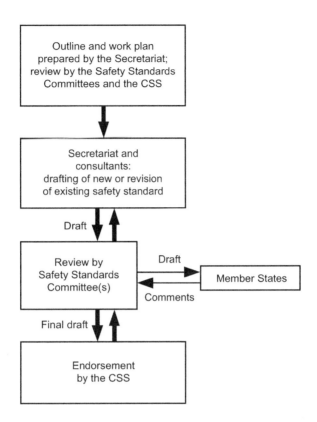

FIG. 2. The process for developing a new safety standard or revising an existing standard.

It articulates the mandate of the IAEA, the vision for the future application of the safety standards, policies and strategies, and corresponding functions and responsibilities.

INTERACTION WITH OTHER INTERNATIONAL ORGANIZATIONS

The findings of the United Nations Scientific Committee on the Effects of Atomic Radiation (UNSCEAR) and the recommendations of international expert bodies, notably the International Commission on Radiological Protection (ICRP), are taken into account in developing the IAEA safety standards. Some safety standards are developed in cooperation with other bodies in the United Nations system or other specialized agencies, including the Food and Agriculture Organization of the United Nations, the United Nations Environment Programme, the International Labour Organization, the OECD Nuclear Energy Agency, the Pan American Health Organization and the World Health Organization.

INTERPRETATION OF THE TEXT

Safety related terms are to be understood as defined in the IAEA Nuclear Safety and Security Glossary (see https://www.iaea.org/resources/publications/iaea-nuclear-safety-and-security-glossary). Otherwise, words are used with the spellings and meanings assigned to them in the latest edition of The Concise Oxford Dictionary. For Safety Guides, the English version of the text is the authoritative version.

The background and context of each standard in the IAEA Safety Standards Series and its objective, scope and structure are explained in Section 1, Introduction, of each publication.

Material for which there is no appropriate place in the body text (e.g. material that is subsidiary to or separate from the body text, is included in support of statements in the body text, or describes methods of calculation, procedures or limits and conditions) may be presented in appendices or annexes.

An appendix, if included, is considered to form an integral part of the safety standard. Material in an appendix has the same status as the body text, and the IAEA assumes authorship of it. Annexes and footnotes to the main text, if included, are used to provide practical examples or additional information or explanation. Annexes and footnotes are not integral parts of the main text. Annex material published by the IAEA is not necessarily issued under its authorship; material under other authorship may be presented in annexes to the safety standards. Extraneous material presented in annexes is excerpted and adapted as necessary to be generally useful.

CONTENTS

1. INTRODUCTION

BACKGROUND

1.1. Requirements for the safety of research reactors, with particular emphasis on their design and operation, are established in IAEA Safety Standards Series No. SSR-3, Safety of Research Reactors [1]. This Safety Guide provides recommendations on the use of a graded approach in the application of these safety requirements.

1.2. For the purpose of this Safety Guide, a graded approach is the application of safety requirements commensurate with the risks associated with a research reactor. The use of a graded approach is intended to ensure that the necessary levels of analysis, documentation and measures are commensurate with the nature and characteristics of a research reactor, the stage in the lifetime of the facility and the magnitude of any radiation risks.

1.3. This Safety Guide was developed together with ten other Safety Guides on the safety of research reactors, as follows:

- IAEA Safety Standards Series No. SSG-80, Commissioning of Research Reactors [2];
- IAEA Safety Standards Series No. SSG-81, Maintenance, Periodic Testing and Inspection of Research Reactors [3];
- IAEA Safety Standards Series No. SSG-82, Core Management and Fuel Handling for Research Reactors [4];
- IAEA Safety Standards Series No. SSG-83, Operational Limits and Conditions and Operating Procedures for Research Reactors [5];
- IAEA Safety Standards Series No. SSG-84, The Operating Organization and the Recruitment, Training and Qualification of Personnel for Research Reactors [6];
- IAEA Safety Standards Series No. SSG-85, Radiation Protection and Radioactive Waste Management in the Design and Operation of Research Reactors [7];
- IAEA Safety Standards Series No. SSG-10 (Rev. 1), Ageing Management for Research Reactors [8];
- IAEA Safety Standards Series No. SSG-37 (Rev. 1), Instrumentation and Control Systems and Software Important to Safety for Research Reactors [9];
- IAEA Safety Standards Series No. SSG-20 (Rev. 1), Safety Assessment of Research Reactors and Preparation of the Safety Analysis Report [10];

— IAEA Safety Standards Series No. SSG-24 (Rev. 1), Safety in the Utilization and Modification of Research Reactors [11].

1.4. The terms used in this Safety Guide, including the definition of a graded approach, are to be understood as defined in the IAEA Nuclear Safety and Security Glossary [12].

1.5. This Safety Guide supersedes IAEA Safety Standards Series No. SSG-22, Use of a Graded Approach in the Application of the Safety Requirements for Research Reactors[1].

OBJECTIVE

1.6. The Safety Guide provides recommendations on the use of a graded approach in the application of the safety requirements established in SSR-3 [1] for research reactors, including critical assemblies and subcritical assemblies, without compromising safety.

1.7. All the safety requirements are applicable to all types of research reactor and cannot be waived (see para. 6.18 of SSR-3 [1]). The recommendations provided in this Safety Guide are on whether and how a graded approach can be applied to specific requirements established in SSR-3 [1].

1.8. This Safety Guide is intended for use by regulatory bodies, operating organizations and other organizations involved in the site evaluation, design, construction, commissioning, operation and preparation for decommissioning of research reactors.

SCOPE

1.9. This Safety Guide considers the application of a graded approach throughout the lifetime of a research reactor (site evaluation, design, construction, commissioning, operation and preparation for decommissioning), including utilization and experiments that are specific features of research reactor operation. A major aspect of this Safety Guide involves the use of a graded approach in the

---

[1] INTERNATIONAL ATOMIC ENERGY AGENCY, Use of a Graded Approach in the Application of the Safety Requirements for Research Reactors, IAEA Safety Standards Series No. SSG-22, IAEA, Vienna (2012).

application of the safety requirements for the design and operation of research reactors, so that the fundamental safety objective (see paras 2.2 and 2.3 of SSR-3 [1]) to protect people and the environment from harmful effects of ionizing radiation is achieved.

1.10. This Safety Guide is primarily intended for use for heterogeneous, thermal spectrum research reactors having a power rating of up to several tens of megawatts. For research reactors of higher power, specialized reactors (e.g. fast spectrum reactors) and reactors having specialized facilities (e.g. hot or cold neutron sources, high pressure and high temperature loops), additional guidance may be needed. Homogeneous reactors and accelerator driven systems are out of the scope of this publication.

1.11. Although this Safety Guide is primarily intended for newly designed and constructed reactors, it may also be applied to existing research reactors to the extent practicable.

STRUCTURE

1.12. Section 2 provides a description of the basic elements of a graded approach and its application. The remaining sections provide recommendations on the application of a graded approach to requirements for regulatory supervision (Section 3); management and verification of safety (Section 4); site evaluation (Section 5); design (Section 6); operation (Section 7); and preparation for decommissioning (Section 8). Section 9 provides recommendations on the interfaces between safety and security. Sections 3–9 have a structure similar to the corresponding sections of SSR-3 [1].

2. BASIC ELEMENTS OF A GRADED APPROACH FOR RESEARCH REACTORS

GENERAL CONSIDERATIONS FOR APPLYING A GRADED APPROACH TO RESEARCH REACTORS

2.1. Requirement 12 of SSR-3 [1] states:

"The use of the graded approach in application of the safety requirements for a research reactor shall be commensurate with the potential hazard of the facility and shall be based on safety analysis and regulatory requirements."

The use of a graded approach in the application of the safety requirements for research reactors in SSR-3 [1] is valid in all stages of the lifetime of a research reactor.

2.2. Paragraph 2.15 of SSR-3 [1] states:

"Research reactors are used for special and varied purposes, such as research, training, education, radioisotope production, neutron radiography and materials testing. These purposes call for different design features and different operational regimes. Design and operating characteristics of research reactors may vary significantly, since the use of experimental devices may affect the performance of reactors. In addition, the need for flexibility in their use requires a different approach to achieving and managing safety."

2.3. Because of the wide range of designs, operating conditions, radioactive inventories and utilization activities, the safety requirements for research reactors established in SSR-3 [1] are not applicable to every research reactor in the same way. For example, the way in which requirements are applied to a multipurpose, high power research reactor might be very different from the way in which the requirements are applied to a research reactor with very low power and very low associated radiation risks to workers, the public and the environment (see para. 2.15 of SSR-3 [1]).

2.4. During the lifetime of a research reactor, the use of a graded approach in the application of the safety requirements should be such that safety functions and operational limits and conditions are preserved, and there are no undue radiation hazards to workers, the public and the environment.

2.5. The use of a graded approach should be based on safety analyses and regulatory requirements and supported by expert judgement. Expert judgement implies that account is taken of the safety functions of structures, systems and components (SSCs) and the consequences of the failure to perform these functions and implies that the judgement is documented and subject to appropriate review and approval using a management system process. Prescriptive regulatory approaches[2], resulting in very detailed regulatory requirements, might restrict the application of a graded approach by the operating organization in respect of certain safety requirements. Other elements to be considered when applying a graded approach are the complexity and the maturity of the technology, operating experience associated with activities and the stage in the lifetime of the research reactor.

DESCRIPTION OF THE USE OF A GRADED APPROACH IN THE APPLICATION OF SAFETY REQUIREMENTS FOR RESEARCH REACTORS

2.6. The result of the use of a graded approach in the application of safety requirements should be a decision on the appropriate effort to be expended and the appropriate manner of complying with each safety requirement, in accordance with the characteristics and the potential hazard of the research reactor.

2.7. The overall method to determine the extent to which a graded approach is applied may be qualitative, quantitative or a combination of both. The approach presented in this Safety Guide has two steps. The first is the qualitative categorization of the research reactor in accordance with its potential hazard (see para. 2.16 of SSR-3 [1]). The second is consideration of a specific safety requirement from SSR-3 [1], and the quantitative and/or qualitative analysis of any activities and/or SSCs associated with that requirement.

---

[2] Prescriptive and performance based regulatory approaches are described in para. 2.80 of IAEA Safety Standards Series No. SSG-16 (Rev. 1), Establishing the Safety Infrastructure for a Nuclear Power Programme [13].

Step 1: Categorization of the research reactor in accordance with potential hazards

2.8. Qualitative categorization of the research reactor should be performed based on the potential radiological hazard, as follows:

(a) Facilities with significant potential for an off-site radiological hazard: such facilities include research reactors with high operating power, a large radioactive inventory or high pressure experimental devices. These facilities are categorized as a high potential hazard.
(b) Facilities with potential for an on-site radiological hazard only: such facilities include research reactors with an operating power up to a few megawatts, a limited radioactive inventory or with no high pressure experimental devices. These facilities are categorized as a medium potential hazard.
(c) Facilities with no potential radiological hazard beyond the research reactor hall and associated beam tubes or connected experimental facility areas: such facilities include facilities with low operating power, not requiring heat removal systems or with a small radioactive inventory. These facilities are categorized as a low potential hazard.

Section 3 of SSG-85 [7] provides further recommendations on evaluating the radiological hazard associated with research reactors.

2.9. Additional characteristics to be considered in categorizing the research reactor in accordance with its potential hazard are listed in para. 2.17 of SSR-3 [1], which states:

"The factors to be considered in deciding whether the application of certain requirements…may be graded include:

(a) The reactor power;
(b) The potential source term;
(c) The amount and enrichment of fissile and fissionable material;
(d) Spent fuel elements, high pressure systems, heating systems and the storage of flammable materials, which may affect the safety of the reactor;
(e) The type of fuel elements;
(f) The type and the mass of moderator, reflector and coolant;
(g) The amount of reactivity that can be introduced and its rate of introduction, reactivity control, and inherent and additional safety features (including those for preventing inadvertent criticality);

(h)   The quality of the containment structure or other means of confinement;
(i)   The utilization of the reactor (experimental devices, tests and reactor physics experiments);
(j)   The site evaluation, including external hazards associated with the site and the proximity to population groups;
(k)   The ease or difficulty in changing [3] the overall configuration."

On the basis of these characteristics, together with the application of expert judgement and consideration of any other factors that might affect the potential hazard, the research reactor should be categorized as a high, medium or low potential hazard.

Step 2: Analysis and application of a graded approach

2.10. Following the categorization of the facility in step 1, an analysis should be performed to determine the appropriate manner for meeting a specific safety requirement using a graded approach. A safety requirement may address a specific SSC or an element of the management system. In such cases, the safety significance of each SSC or management system element (including SSCs and elements related to experiments) can be determined through the step 2 analysis. Requirement 16 of SSR-3 [1] states that "**All items important to safety for a research reactor facility shall be identified and shall be classified on the basis of their safety function and their safety significance.**"

2.11. In this step, the level of detail at which requirements are applied to activities and/or SSCs is determined, in accordance with the importance to safety of the activity or SSC. The level of detail should cover, for example, the rigour of the analysis to be conducted, the frequency of activities such as testing and preventive maintenance, the stringency of required approvals and the degree of oversight of activities.

2.12. The safety significance of SSCs should be determined by conducting a safety assessment (see SSG-20 (Rev. 1) [10]) analysing the consequences of a failure of the intended safety function to be performed by the considered SSCs. Based on the safety class, appropriate design requirements should be assigned to meet para. 6.32 of SSR-3 [1].

---

[3]   Modifications and experiments are an important aspect of research reactor design and operation. See paras 6.140–6.142 and 7.70 for specific recommendations.

2.13. With regard to analysing the safety significance of elements of the management system, and then applying a graded approach to meeting the safety requirements for these elements, para. 4.15 of IAEA Safety Standards Series No. GSR Part 2, Leadership and Management for Safety [14] states:

> "The criteria used to grade the development and application of the management system shall be documented in the management system. The following shall be taken into account:
>
> (a) The safety significance and complexity of the organization, operation of the facility or conduct of the activity;
> (b) The hazards and the magnitude of the potential impacts (risks) associated with the safety, health, environmental, security, quality and economic elements of each facility or activity [15–18];
> (c) The possible consequences for safety if a failure or an unanticipated event occurs or if an activity is inadequately planned or improperly carried out."

Paragraphs 2.37–2.40 of IAEA Safety Standards Series No. GS-G-3.1, Application of the Management System for Facilities and Activities [19], provide recommendations on how elements of the management system can be assessed to support a graded approach in the application of management system requirements.

2.14. The analysis in step 2 to determine how requirements related to SSCs and management system elements are met should consider the overall categorization of the facility from step 1, and the safety significance of the SSC or management system element that is affected. From this analysis, the appropriate level of effort needed in meeting the requirement, and the manner in which the requirement will be met, should be determined. The insight from expert judgement (from a single expert or from a multidisciplinary group, as appropriate) may be introduced into the decision making process after the results of this analysis are available.

2.15. Specific recommendations on the use of a graded approach in the application of each safety requirement of SSR-3 [1] are provided in Sections 3–9. Examples are given for the graded application of requirements for research reactors with a high, medium or low potential hazard.

3. USE OF A GRADED APPROACH IN THE REGULATORY SUPERVISION OF RESEARCH REACTORS

3.1. General requirements for the legal and regulatory infrastructure for facilities and activities are established in IAEA Safety Standards Series No. GSR Part 1 (Rev. 1), Governmental, Legal and Regulatory Framework for Safety [20], which includes requirements on the use of a graded approach in relation to the responsibilities and functions of the regulatory body. IAEA Safety Standards Series No. GSG-13, Functions and Processes of the Regulatory Body for Safety [21] provides recommendations on the core regulatory functions and processes, including the application of a graded approach (see section 2 of GSG-13 [21]), to the following:

(a) Regulations and guides;
(b) Notification and authorization;
(c) Review and assessment of facilities and activities;
(d) Inspection of facilities and activities;
(e) Enforcement;
(f) Emergency preparedness and response;
(g) Communication and consultation with interested parties.

THE USE OF A GRADED APPROACH IN THE LEGAL AND REGULATORY INFRASTRUCTURE FOR RESEARCH REACTORS

3.2. The requirements established in GSR Part 1 (Rev. 1) [20] for the legal infrastructure are placed on the government (e.g. for the adoption of legislation that assigns the prime responsibility for safety to the operating organization and establishes a regulatory body) and on the regulatory body (e.g. for the establishment of regulations that results in a system of authorization for the regulatory control of facilities and activities and for the enforcement of the regulations). Regarding the application of these requirements, para. 3.2 of SSR-3 [1] states that "The application of a graded approach that is commensurate with the potential hazards of the facility is essential and shall be used in the determination and application of adequate safety requirements".

3.3. Specific aspects of the legal and regulatory framework in a State may affect the extent to which a graded approach can be used. In a State where the most hazardous nuclear installation is a single research reactor with a low potential hazard (see para. 2.8), for the implementation of the national policy and strategy for safety, a graded approach can be used, with a less comprehensive set of policy mechanisms and internal resources than in a State with a large and diverse nuclear infrastructure. A graded approach to applying the requirements for a State's legal and regulatory infrastructure[4] should include an analysis of the radiation risks associated with facilities and activities and also consider the following provisions that are necessary for the government to meet the fundamental safety objective:

(a) Human and financial resources;
(b) The type of authorization process;
(c) The provisions for regulatory review;
(d) Appropriate inspection and enforcement regulations;
(e) Communication and consultation with interested parties.

See also Requirements 1 and 2 of GSR Part 1 (Rev. 1) [20] on the establishment of a national policy, strategy and framework for safety.

THE USE OF A GRADED APPROACH IN THE ORGANIZATION
AND FUNCTIONS OF THE REGULATORY BODY
FOR RESEARCH REACTORS

3.4. Paragraph 4.3 of GSR Part 1 (Rev. 1) [20] states:

"The objective of regulatory functions is the verification and assessment of safety in compliance with regulatory requirements. The performance of regulatory functions shall be commensurate with the radiation risks associated with facilities and activities, in accordance with a graded approach."

The performance of regulatory functions should be commensurate with the categorization of research reactors, as described in para. 2.8.

3.5. The regulatory body is required to be provided with sufficient authority, and a sufficient number of experienced staff and financial resources to discharge its assigned responsibilities (see Requirement 3 of GSR Part 1 (Rev. 1) [20]). The

---

[4] Some examples of the application of a graded approach to a national legal and regulatory infrastructure are shown in Ref. [22].

responsibilities of the regulatory body include establishing regulations, reviewing and assessing safety related information (e.g. from the safety analysis report for a research reactor; see Requirement 1 of SSR-3 [1]), issuing authorizations, performing inspections, taking enforcement actions, and providing information to other competent authorities and the public.

3.6. Examples of safety requirements for the regulatory body that can be met using a graded approach are requirements for: staffing; resources for in-house technical support; inspections; the content and detail of authorizations, regulations and guides; and the detail required from the licensee for submissions of documentation on the safety of the facility, including the safety analysis report (see section 4 of GSR Part 1 (Rev. 1) [20]). Recommendations on the application of a graded approach to the organization of the regulatory body and to the performance of regulatory functions are provided in IAEA Safety Standards Series No. GSG-12, Organization, Management and Staffing of the Regulatory Body for Safety [23] and GSG-13 [21], respectively. Regulatory requirements should also consider the potential for limiting facilities' ability to apply a graded approach to these requirements and the scope of a graded approach in the application of requirements for the regulatory body itself.

THE USE OF A GRADED APPROACH IN THE AUTHORIZATION PROCESS FOR RESEARCH REACTORS

3.7. The authorization process is often performed in steps for the various stages of the lifetime of a research reactor. Paragraph 3.4 of SSR-3 [1] states (footnote omitted):

"The authorization process may vary among States, but the major stages of the authorization process for nuclear research reactors shall include the following:

(a) Site evaluation;
(b) Design;
(c) Construction;
(d) Commissioning;
(e) Operation, including utilization and modification;
(f) Decommissioning;
(g) Release from regulatory control."

At each of these stages, regulatory reviews and assessments are usually made and authorizations or approvals are issued. In some cases, some of these stages may be combined, depending on the nature of the research reactor and the regulatory requirements.

3.8. The authorization process is used by the regulatory body to exercise control during all stages of the lifetime of the research reactor. This control is accomplished by means of the following:

(a) Defining clear lines of authority for authorizations to proceed;
(b) Reviewing and assessing all safety relevant documents, particularly the safety analysis report;
(c) Issuing of permits and licences, for the various stages;
(d) Implementing hold points for inspections, review and assessment;
(e) Reviewing, assessing and approving operational limits and conditions;
(f) Authorizing construction;
(g) Authorizing commissioning;
(h) Authorizing operation;
(i) Authorizing operating personnel;
(j) Authorizing decommissioning.

3.9. The steps in the authorization process apply to all research reactors at all stages of their lifetime and should apply to experiments and modifications depending on their importance to safety (see SSG-24 (Rev. 1) [11]). However, at each step in the authorization process, a graded approach may be used in the application of the safety requirements by the regulatory body, depending on the potential hazard associated with the stage in the lifetime of the research reactor or with the experiment or modification, as appropriate. This may include, for example, the level of detail needed in the application for an authorization, the depth of review and the resources deployed by the regulatory body when considering an application for authorization, and the duration of an authorization when it is issued.

Safety analysis report

3.10. Requirement 1 of SSR-3 [1] states:

> **"A safety analysis report shall be prepared by the operating organization for a research reactor facility. The safety analysis report shall provide a justification of the site and the design and shall provide a basis for the safe operation of the research reactor. The safety analysis report shall**

be reviewed and assessed by the regulatory body before the research reactor project is authorized to progress to the next stage. The safety analysis report shall be periodically updated over the research reactor's operating lifetime to reflect modifications made to the facility and on the basis of experience and in accordance with regulatory requirements."

3.11. The responsibilities of the regulatory body include the review and assessment of safety related information from the safety analysis report as part of the authorization process for the research reactor (see paras 3.5, 3.10 and 3.11 of SSR-3 [1]). A graded approach is required to be applied to this review and assessment (see Requirement 26 of GSR Part 1 (Rev. 1) [16] and para. 3.10 of SSR-3 [1]).

3.12. Paragraph 3.8 of SSR-3 [1] states:

"The level of detail of the information to be presented in the safety analysis report shall be determined using a graded approach. For reactors with high power levels, the safety analysis report will usually require more detail in discussions such as those of reactor design and accident scenarios. For some reactors (e.g. research reactors with a low potential hazard, critical or subcritical assemblies), the requirements for the safety analysis report content may be much less extensive."

The level of detail necessary to demonstrate that acceptance criteria are met should be commensurate with the potential hazard of the research reactor. For a facility with a low potential hazard, the safety analysis may include bounding analyses, owing to large safety margins in the design, to demonstrate that the research reactor can be operated safely. For research reactors with a higher potential hazard, typically, more detailed analysis is necessary to demonstrate safety in operational states and in accident conditions, with less use of large bounding analyses.

3.13. The complementary probabilistic safety assessment which might be carried out to supplement deterministic safety analysis, if appropriate (see Requirement 5 of SSR-3 [1]), is another element of the safety analysis report requirement that could vary in accordance with the potential hazard of the facility. The appendix to SSG-20 (Rev. 1) [10] provides recommendations on safety assessment and the safety analysis report for research reactors, including the application of a graded approach commensurate with the magnitude of the potential hazards.

THE USE OF A GRADED APPROACH IN INSPECTION AND ENFORCEMENT FOR RESEARCH REACTORS

3.14. General safety requirements for inspection and enforcement are established in Requirements 27–31 of GSR Part 1 (Rev. 1) [20], and specific requirements for research reactors are established in paras 3.13–3.16 of SSR-3 [1]. With regard to applying a graded approach to inspections, para. 4.50 of GSR Part 1 (Rev. 1) [20] states:

"The regulatory body shall develop and implement a programme of inspection of facilities and activities, to confirm compliance with regulatory requirements and with any conditions specified in the authorization. In this programme, it shall specify the types of regulatory inspection (including scheduled inspections and unannounced inspections), and shall stipulate the frequency of inspections and the areas and programmes to be inspected, in accordance with a graded approach."

In general, there may be fewer inspections and hold points for a research reactor with a low potential hazard, compared with those for a research reactor with a higher potential hazard.

3.15. Enforcement actions are required to be commensurate with the significance for safety of the non-compliance (see para. 4.54 of GSR Part 1 (Rev. 1) [20]). Regulatory bodies should allocate resources and apply enforcement actions or methods in a manner commensurate with the seriousness of the non-compliance, increasing them as necessary to bring about compliance with requirements.

3.16. Factors that should be considered in determining the appropriate level of enforcement actions are as follows (see also para. 3.308 of GSG-13 [21]):

(a)   The safety significance of the non-compliance or of the violation of regulatory requirements;
(b)   Whether the non-compliance or violation is repeated;
(c)   Whether there has been an intentional violation;
(d)   Whether or not the authorized party identified and/or reported the non-compliance or the violation;
(e)   Whether the non-compliance or violation impacted the ability of the regulatory body to perform its regulatory oversight function;

(f) The past safety performance of the authorized party and the performance trend (noting that past good performance does not ease the enforcement imposed);

(g) The need for consistency and openness in the treatment of authorized parties.

3.17. Enforcement actions in response to an intentional violation of a regulatory requirement should be commensurately serious.

4. USE OF A GRADED APPROACH IN THE MANAGEMENT AND VERIFICATION OF SAFETY OF RESEARCH REACTORS

4.1. Requirements for the management system for operating organizations of facilities, including research reactors, are established in GSR Part 2 [14]. Requirement 7 of GSR Part 2 [14] states that "**The management system shall be developed and applied using a graded approach.**" Additional requirements specific to the management system for research reactors are established in Requirements 2–6 of SSR-3 [1].

A GRADED APPROACH TO MANAGEMENT RESPONSIBILITIES FOR SAFETY FOR RESEARCH REACTORS

4.2. Requirements for responsibilities in the management of safety for research reactors are established in Requirement 2 of SSR-3 [1]. Paragraph 4.1 of SSR-3 [1] states (footnote omitted):

"In order to ensure rigour and thoroughness at all levels of the staff in the achievement and maintenance of safety, the operating organization:

(a) Shall establish and implement safety policies and shall ensure that safety matters are given the highest priority;

(b) Shall clearly define responsibilities and accountabilities with corresponding lines of authority and communication;

(c) Shall ensure that it has sufficient staff with appropriate qualifications and training at all levels;

(d) Shall develop and strictly adhere to sound procedures for all activities that may affect safety, ensuring that managers and supervisors promote and support good safety practices, while correcting poor safety practices;

(e) Shall review, monitor and audit all safety related matters on a regular basis, and shall take appropriate corrective actions where necessary;

(f) Shall develop and sustain a strong safety culture, and shall prepare a statement of safety policy and safety objectives, which is disseminated to and understood by all staff."

There are elements of this requirement that cannot be graded, for example, for the operating organization to have prime responsibility for the safety of the research reactor, and the requirement to develop and sustain a strong culture for safety.

4.3. The resources allocated to the management of a research reactor should vary depending on the potential hazard of the facility, its complexity and its size. For example, in a research reactor with a high potential hazard, the requirement for sufficient staff (see para. 7.14 of SSR-3 [1]) may result in a large operating organization to enable continuous operation day and night, and to provide maintenance and technical support. In facilities such as some low potential hazard research reactors and critical and subcritical assemblies, the requirement for sufficient staff may result in a small operating organization, albeit still with the necessary training to operate, maintain and ensure the safety of the research reactor. The organizational structure for the operating organization and the minimum staffing are also required to take into account the personnel needed to respond to accident conditions (see para. 7.14 of SSR-3 [1]).

SAFETY POLICY FOR RESEARCH REACTORS

4.4. Requirement 3 of SSR-3 [1] states that "**The operating organization for a research reactor facility shall establish and implement safety policies that give safety the highest priority.**"

4.5. The way this requirement to establish and implement a safety policy applies is the same irrespective of the potential hazard of the facility. The safety policy is a central component of the management system for any nuclear facility, to ensure that all activities within the operating organization give safety the highest priority.

THE USE OF A GRADED APPROACH IN THE APPLICATION OF THE REQUIREMENTS FOR THE MANAGEMENT SYSTEM FOR A RESEARCH REACTOR

4.6. Requirements for the management system for a research reactor are established in Requirement 4 of SSR-3 [1]. According to para. 4.7 of SSR-3 [1], the complexity of the management system for a particular research reactor and associated experimental facilities is required to be commensurate with the potential hazard of the reactor and experimental facilities; it should also meet the requirements of the regulatory body. Requirements for the preparation and implementation of a graded management system are established in Requirement 7 and para. 4.15 of GSR Part 2 [14].

4.7. In general, management system processes should be more stringent for items and services where a failure or a non-conformance has the highest potential hazard. For other items and services, the management system processes may be less stringent. The following are examples of elements of the management system where this requirement can be applied using a graded approach:

(a) Type, duration and content of training;
(b) Level of detail and degree of review and approval of operating procedures;
(c) Need for and detail of inspection plans;
(d) Scope, depth and frequency of operational safety reviews and controls, including internal and independent audits;
(e) Type and frequency of safety assessments;
(f) Records to be generated and retained;
(g) Reporting level and authorities of non-conformances and corrective actions;
(h) Maintenance, periodic testing and inspection activities;
(i) Equipment to be included in plant configuration control;
(j) Control applied to the storage and records of spare parts;
(k) Need to analyse events and equipment failure data.

4.8. Operating procedures for a research reactor (see Requirement 74 of SSR-3 [1]) should be subject to a level of review and approval commensurate with their safety significance. A procedure for a simple maintenance task on a component in a non-active system with low safety significance could be written by an experienced member of the operating personnel and reviewed by a maintenance supervisor. A procedure for use in the control room to start up the reactor should be subject to more rigour in the level of detail and the extent of review. For a research reactor with a low potential hazard, the expertise necessary to write and review new procedures might not always exist within the operating organization and might involve experts from the reactor designer or another external organization with appropriate knowledge. The level of review for procedures should also be commensurate with their safety significance.

4.9. The approval of procedures is the responsibility of the reactor manager (see paras 5.14–5.18 of SSG-83 [5]). In every research reactor, regardless of potential hazard, every procedure in the management system should be periodically reviewed by the reactor manager or a designate to enable improvements to be identified.

4.10. Paragraphs 2.37–2.44 of GS-G-3.1 [19] also provide recommendations on a graded approach to the application of requirements for management system controls.

4.11. The requirements in para. 4.20 of SSR-3 [1] for the assessment and improvement of the management system for a research reactor can be applied using a graded approach to identify and correct weaknesses in accordance with their safety significance, and with the potential hazard of the facility. For example, for a research reactor with a high potential hazard, the operating organization may be large, and the management system could include a large number of processes to ensure that operation, utilization and maintenance activities are conducted safely. A process should be implemented by a small group of personnel within the operating organization to identify weaknesses and improvements in the management system on a weekly basis for the reactor management to set priorities based on safety significance. In parallel, the management system should be the subject of periodic external assessment, to identify where improvements can be made. For a research reactor with a low potential hazard, the management system could consist of relatively few processes and procedures, and an audit of the management system could occur as part of the periodic safety review or the renewal of the authorization from the regulatory body.

THE USE OF A GRADED APPROACH IN THE VERIFICATION OF SAFETY AT A RESEARCH REACTOR

Safety assessment

4.12. Requirements for safety assessments to verify the adequacy of the design of the research reactor are established in Requirement 5 of SSR-3 [1]. These requirements can be applied using a graded approach, for example, by taking the potential hazard of the research reactor into account when determining the frequency and scope of safety assessments (such as self-assessments, independent assessment and peer reviews) throughout the lifetime of the facility. The frequency and scope of these assessments should be commensurate with the potential hazard of the facility, recent operating experience, the potential hazard associated with modifications to the research reactor (see paras 7.70–7.74 of this Safety Guide), or the results from previous periodic safety reviews (see para. 4.25 of SSR-3 [1]).

4.13. The requirement to verify the adequacy of the design using safety assessment techniques can be applied using a graded approach based on the potential hazard of the facility and the number of SSCs important to safety, as discussed in para. 6.6. Further recommendations on the use of a graded approach in the safety analysis of the design of a research reactor are provided in paras 6.78–6.84.

Reactor safety committee

4.14. Requirement 6 of SSR-3 [1] states that "**A safety committee (or an advisory group) that is independent from the reactor manager shall be established to advise the operating organization on all the safety aspects of the research reactor.**" The main element of this requirement cannot be graded (i.e. the establishment of a reactor safety committee is required for all research reactors). A minimum list of items that the reactor safety committee is required to review is provided in para. 4.27 of SSR-3 [1] (see also paras 7.9 and 7.10 of this Safety Guide).

4.15. Some aspects of this requirement can be applied using a graded approach; these include the number, size and frequency of reactor safety committee meetings, and the membership of the committee.

4.16. In a research reactor with a high potential hazard, the reactor safety committee may have a busy schedule of work, involving frequent meetings to review proposed experiments of safety significance, safety documentation, reports on radiation doses received by personnel, and reports to the regulatory body. In such a research reactor, the reactor safety committee may designate subcommittees with specific expertise to provide advice or recommendations on specific technical areas such as criticality safety or radiation protection, to reduce the workload on other reactor safety committee members. The composition of the reactor safety committee (and any subcommittees) should include expertise in all technical areas of operation. The operating organization for a research reactor with a high potential hazard typically can form the reactor safety committee using internal personnel. In a research reactor with a low potential hazard, the reactor safety committee may meet less frequently to review the safety of the research reactor and to provide advice to the reactor manager, with additional meetings arranged only as necessary. The operating organization for such a research reactor is typically smaller in size, and the reactor safety committee could include a number of external persons with experience from other facilities in appropriate technical areas.

5. THE USE OF A GRADED APPROACH IN SITE EVALUATION FOR RESEARCH REACTORS

5.1. Requirements for site evaluation for research reactors are established in IAEA Safety Standards Series No. SSR-1, Site Evaluation for Nuclear Installations [24]. Requirement 3 of SSR-1 [24] addresses the scope of site evaluation, including the application of a graded approach to facilities other than nuclear power plants. Recommendations on the application of those requirements for research reactors, using a graded approach, are provided in section 6 of IAEA Safety Standards Series No. SSG-35, Site Survey and Site Selection for Nuclear Installations [25].

5.2. Paragraph 5.1 of SSR-3 [1] states:

"The main safety objective in evaluating the site for a research reactor is the protection of the public and the protection of the environment against the radiological consequences of normal and accidental releases of radioactive material".

To meet this requirement, it is necessary to assess those characteristics of the site that could affect the safety of the research reactor, to determine whether there are deficiencies in the site and if they can be mitigated by appropriate design features, site protection measures and/or administrative procedures. To apply a graded approach to site evaluation, the scope and depth of site evaluation studies and evaluations should be commensurate with the potential radiation risk associated with the research reactor. The scope and detail of the site evaluation may also be reduced if the operating organization adopts conservative parameters for design purposes that reduce the potential for on-site and off-site consequences in the event of an accident; this may be the preferred approach for research reactors. For example, using a conservative design for a particular SSC, which is readily accommodated in the overall design, may simplify the site evaluation.

5.3.   Paragraphs 4.1–4.5 of SSR-1 [24] develop the basis for applying a graded approach to the site evaluation process, to ensure that it is commensurate with the potential hazard of the research reactor. Paragraph 4.5 of SSR-1 [24] states:

"For site evaluation for nuclear installations other than nuclear power plants, the following shall be taken into consideration in the application of a graded approach:

(a)   The amount, type and status of the radioactive inventory at the site (e.g. whether the radioactive material on the site is in solid, liquid and/or gaseous form, and whether the radioactive material is being processed in the nuclear installation or is being stored on the site);

(b)   The intrinsic hazards associated with the physical and chemical processes that take place at the nuclear installation;

(c)   For research reactors, the thermal power;

(d)   The distribution and location of radioactive sources in the nuclear installation;

(e)   The configuration and layout of installations designed for experiments, and how these might change in future;

(f)   The need for active systems and/or operator actions for the prevention of accidents and for the mitigation of the consequences of accidents;

(g)   The potential for on-site and off-site consequences in the event of an accident."

5.4.   Section 9 of IAEA Safety Standards Series No. SSG-9 (Rev. 1), Seismic Hazards in Site Evaluation for Nuclear Installations [26] provides recommendations on a graded approach to the application of Requirement 15 of SSR-1 [24] for the evaluation of the seismic hazard for nuclear installations other than nuclear power plants. The approach is based upon the complexity of the installation and the potential radiological hazards, including hazards due to other materials (e.g. the presence of flammable, explosive or toxic materials). A seismic hazard assessment should initially apply a conservative screening process in which it is assumed that the entire radioactive inventory of the research reactor is released by an accident initiated by a seismic event. If such a release would not lead to unacceptable consequences for workers, the public or the environment, the research reactor may be screened out from further seismic hazard assessment. If the results of the conservative screening process show that the potential consequences of such a release could be significant, an evaluation of seismic hazard is required to be performed in accordance with section 5 of SSR-3 [1] and para. 9.7 of SSG-9 (Rev. 1) [26].

5.5. Section 7 of IAEA Safety Standards Series No. SSG-21, Volcanic Hazards in Site Evaluation for Nuclear Installations [27] provides recommendations on a graded approach to the application of Requirement 17 of SSR-1 [24] with respect to volcanic hazards in site evaluation. A volcanic hazard assessment should initially apply a conservative screening process in which it is assumed that the entire radioactive inventory of the installation is released by an accident initiated by a volcanic event. If such a release would not lead to unacceptable consequences for workers, the public or the environment, the installation may be screened out from further volcanic hazard assessment. If the results of the conservative screening process show that the potential consequences of such a release could be significant, a more detailed volcanic hazard assessment should be performed in accordance with section 5 of SSR-3 [1], using the graded approach recommended in SSG-21 [27] to categorize the installation for the purposes of volcanic hazard assessment.

5.6. Recommendations on a graded approach to the application of Requirements 18–20 of SSR-1 [24] on the evaluation of meteorological and hydrological hazards in site evaluation are provided in IAEA Safety Standards Series No. SSG-18, Meteorological and Hydrological Hazards in Site Evaluation for Nuclear Installations [28]. For the purpose of evaluating meteorological and hydrological hazards, including flooding, the site should be screened on the basis of the complexity of the research reactor, the potential radiological hazards and the hazards due to other materials (e.g. the presence of flammable, explosive or toxic materials). If the results of a conservative screening process show that the consequences of a potential release could be significant, a detailed meteorological and hydrological hazard assessment for the research reactor should be performed in accordance with section 5 of SSR-3 [1], using the graded approach recommended in section 10 of SSG-18 [28].

5.7. Human induced events cannot be included in site evaluation using the same approach as other external events. Because human induced events are discrete and are not characterized by a range of frequency and severity, only one intensity level for each event is expected to be considered in the design basis. Recommendations on the screening and analysis of hazards associated with human induced events are provided in IAEA Safety Standards Series No. SSG-79, Hazards Associated with Human Induced External Events in Site Evaluation for Nuclear Installations [29].

6. THE USE OF A GRADED APPROACH IN THE DESIGN OF RESEARCH REACTORS

6.1. Section 6 of SSR-3 [1] establishes requirements for the design of research reactors under three categories:

(a) Principal technical requirements: Paragraphs 6.2–6.16 of this Safety Guide provide recommendations on the use of a graded approach in the application of Requirements 7–15 of SSR-3 [1].
(b) General requirements for the design: Paragraphs 6.17–6.84 of this Safety Guide provide recommendations on the use of a graded approach in the application of Requirements 16–41 of SSR-3 [1].
(c) Specific requirements for the design: Paragraphs 6.85–6.142 of this Safety Guide provide recommendations on the use of a graded approach in the application of Requirements 42–66 of SSR-3 [1].

THE USE OF A GRADED APPROACH IN THE PRINCIPAL TECHNICAL REQUIREMENTS FOR THE DESIGN OF RESEARCH REACTORS

Main safety functions

6.2. Requirement 7 of SSR-3 [1] states:

> "**The design for a research reactor facility shall ensure the fulfilment of the following main safety functions for the research reactor for all states of the facility: (i) control of reactivity; (ii) removal of heat from the reactor and from the fuel storage; and (iii) confinement of the radioactive material, shielding against radiation and control of planned radioactive releases, as well as limitation of accidental radioactive releases.**"

The use of a graded approach should result in design features that fully meet this requirement and are appropriate for the potential hazard from the research reactor. The control of radioactive discharges (see Requirements 59 and 64 of SSR-3 [1]) is necessary to protect the public and the environment and to meet regulatory requirements, and the way this requirement is applied cannot be graded.

6.3. A graded approach can be used in the application of some elements of Requirement 7 of SSR-3 [1] for the main safety functions, as follows:

(a) Control of reactivity:
   (i)   The capability to shut down the reactor when necessary is a requirement for all research reactors, although the size of the subcriticality margin available and the speed of response of the shutdown system may vary in accordance with the reactor design.
   (ii) Some research reactors may have inherent self-limiting power levels and/or systems that physically limit the amount of positive reactivity that can be inserted into the core. This property can be used for a graded approach in the design of the shutdown system.
(b) Removal of heat from the reactor and from the fuel storage:
   (i)   For some research reactors (typically with a medium or high potential hazard and higher power) a forced convection cooling system to remove fission heat could be necessary to meet the acceptance criteria for the design, in all operational states and in accident conditions, whereas for research reactors with less demanding cooling needs, such as some critical assemblies and subcritical assemblies, fission heat could be generated at sufficiently low levels that it could be adequately removed without the need for an engineered system.
   (ii) Similarly, for the removal of decay heat following shutdown, in the design of the cooling system, a graded approach can be used, based on factors such as the power of the research reactor, the maximum level of fission products and the heat transfer characteristics of the fuel. For a research reactor with less demanding cooling needs, where no heat removal system is necessary during operation, no dedicated equipment is necessary for decay heat removal.
   (iii) The scope and necessity of coolant systems (see Requirement 47 of SSR-3 [1]), including emergency core cooling systems to make up the inventory of reactor coolant in the event of a loss of coolant accident (see Requirement 48 of SSR-3 [1]), should be verified through the safety analysis for the research reactor, which is required to demonstrate that for all operational states and accident conditions, the main safety function of heat removal is fulfilled.
(c) Confinement of radioactive material, shielding against radiation and control of planned radioactive releases, as well as limitation of accidental radioactive releases:
   (i)   In the design of SSCs to perform barrier or retention functions to confine radioactive material in operational states and in accident conditions, a graded approach can be used. The approach can be based on the potential hazard of the research reactor, the inventory of fission products, the characteristics of the fuel and the results of the safety

analysis for the research reactor (see also the description of the fourth level of defence in depth in para. 6.8).

(ii) The design of shielding against radiation should be based on the magnitude of the radiation hazard calculated for each location in the research reactor where actions by operating personnel are necessary in operational states and in accident conditions, and for appropriate locations outside the research reactor. The appropriate material and thickness of shielding that is commensurate with the hazard can then be included in the design.

(iii) The control of planned radioactive discharges is required for all research reactors regardless of their potential hazard (see para. 6.2).

Radiation protection

6.4. Requirements for radiation protection in the design of research reactors are established in Requirement 8 of SSR-3 [1]. The requirement for the design to ensure that doses to reactor personnel and the public do not exceed the established dose limits and are kept as low as reasonably achievable inherently implies the use of a graded approach in which the potential hazard of the research reactor and its characteristics, such as the inventory of fission products and the proximity to a population centre, are also taken into account. Specific design provisions, or SSCs included in the design to protect reactor personnel and the public from radiation (e.g. an emergency filtration system), could be larger and/or more complex for a research reactor with a higher potential hazard.

Design

6.5. Requirements for the design of a research reactor are established in Requirement 9 of SSR-3 [1]. The use of a graded approach in the application of this requirement should be based on the potential hazard of the research reactor and the factors listed in para. 2.9.

6.6. Paragraph 6.9 of SSR-3 [1] requires that adequate information on the design is available for operation, future modifications and decommissioning; this requirement can be applied using a graded approach based on the potential hazard of the research reactor, the number of SSCs important to safety and the number of SSCs in the research reactor that have associated radiation hazards. The quantity of information that would be adequate to decommission a research reactor with a high potential hazard should be larger in scope than for research reactors with a lower potential hazard (e.g. some low power research reactors, critical assemblies, subcritical assemblies).

Application of the concept of defence in depth

6.7.   Requirements for the application of the concept of defence in depth to the design of a research reactor are established in Requirement 10 of SSR-3 [1]. Paragraph 2.12 of SSR-3 [1] describes the five levels of defence in depth for preventing or controlling deviations in normal operation, for preventing accidents and for mitigating the radiological consequences of accidents.

6.8.   Defence in depth is an important design principle that is required for all research reactors regardless of the potential hazard. The first four levels of defence in depth should be included in the design. In the design capability of the engineered safety features, a graded approach can be used. For a facility with a low or medium potential hazard or a critical or subcritical assembly, for example, the decay heat load could be smaller, and typically a smaller fission product inventory needs to be confined or mitigated than for a research reactor with a high potential hazard. It should be recognized that for low potential hazard research reactors, and some critical assemblies and subcritical assemblies, the types of accident that the fourth or fifth level of defence in depth are intended to cope with might not be physically possible.

Interfaces of safety with security and the State system of accounting for, and control of, nuclear material

6.9.   Requirement 11 of SSR-3 [1] states:

> **"Safety measures, nuclear security measures and arrangements for the State system of accounting for, and control of, nuclear material for a research reactor shall be designed and implemented in an integrated manner so that they do not compromise one another."**

This requirement is specifically for the integration of measures and arrangements, and consequently the way it is applied cannot be graded.[5]

Proven engineering practices

6.10. Requirement 13 of SSR-3 [1] states that **"Items important to safety for a research reactor shall be designed in accordance with the relevant national and international codes and standards."**

---

[5] Practical guidance on this topic is provided in Ref. [30].

6.11. For SSCs for which there are no established codes or standards, para. 6.22 of SSR-3 [1] allows the use of related standards or the results of experience, tests or analysis, and requires that such an approach be justified. A graded approach can be used in the application of this requirement, based on the potential hazard of the facility, the safety classification of the SSC, and the availability of related codes and standards, such as those for nuclear power plants or from other industries. Expert judgement is necessary in using this approach and this should be documented as part of the written justification, which should be approved in accordance with a process in the management system.

6.12. If the design process does not follow established engineering practice, para. 6.23 of SSR-3 [1] requires that "a process shall be established under the management system to ensure that safety is demonstrated". A graded approach can be used in the application of this requirement based on the safety classification of the SSC, its reliability requirements and the consequences of failure established in the safety analysis. The effort needed to develop the new process and its scope and level of detail should be commensurate with the hazard category of the research reactor and the safety classification of the SSC. In all cases, para. 6.23 of SSR-3 [1] requires that the SSC be monitored in service to verify that the research reactor operates as designed.

Provision for construction

6.13. Requirement 14 of SSR-3 [1] states:

> **"Items important to safety for a research reactor facility shall be designed so that they can be manufactured, constructed, assembled, installed and erected in accordance with established processes that ensure the achievement of the design specifications and the required level of safety."**

The way this requirement for items important to safety to perform in accordance with design specifications is applied cannot be graded, and the ability of those SSCs to function as designed cannot be compromised by the manufacturing, construction and installation processes.

Features to facilitate radioactive waste management and decommissioning

6.14. Requirements for features to facilitate radioactive waste management and decommissioning are established in Requirement 15 of SSR-3 [1] and can be applied using a graded approach.

6.15. The choice of materials used in the design of a research reactor should involve engineering judgement to balance the utilization needs of the facility with the needs associated with waste management and the hazards in the decommissioning process that result from long lived activation products. The effort and scope of design measures to minimize radioactive waste generated during operation or decommissioning of the research reactor, and to manage waste that is generated, should be commensurate with the potential hazard of the research reactor and the potential for generation of activation products. For a research reactor with a high potential hazard, the elimination of materials that produce long lived activation products might not be feasible; however, minimizing them where possible will reduce the overall potential hazard for the decommissioning process. Planning for how activated materials or waste are managed during the operating lifetime and the decommissioning of the research reactor should include radiation protection considerations and could include specific technology or practices to prevent undue radiation exposure of personnel. For example, it may be necessary to include special processing and storage facilities to manage waste generated during operation. Also, the facility could be designed so that any highly activated materials can be easily assessed and removed during decommissioning, to minimize exposure. For a research reactor with a low potential hazard, such as some critical and subcritical assemblies, there might not be a significant hazard from activation products.

6.16. Requirement 15 of SSR-3 [1] also applies to modifications and new experiments undertaken during the operation of the research reactor. In such cases, a graded approach could be applied to the choice of material used in the design of new experimental equipment based on the potential hazard introduced in relation to waste management and decommissioning.

THE USE OF A GRADED APPROACH IN THE GENERAL REQUIREMENTS FOR THE DESIGN OF RESEARCH REACTORS

Safety classification of structures, systems and components

6.17. Requirements for the safety classification of SSCs are established in Requirement 16 of SSR-3 [1]. The classification of SSCs important to safety is useful input when using a graded approach in the application of other requirements.

6.18. SSCs important to safety at all research reactors, regardless of the potential hazard, are required to be classified. The method for determining the safety significance of SSCs is required to be based on deterministic methods, complemented by probabilistic methods, and may be supported by engineering judgement (see para. 6.29 of SSR-3 [1]). Research reactors with higher potential hazard and significant in-core experimental facilities, such as loops, typically include a greater number of SSCs that are in a higher safety class.

Design basis for items important to safety

6.19. Requirements for the design basis for items important to safety are established in Requirement 17 of SSR-3 [1]. The requirement to justify and document the design basis for each item important to safety in para. 6.33 of SSR-3 [1] can be applied using a graded approach based on the potential hazard of the research reactor and the level of detail for each SSC that is needed to enable the operating organization to operate the research reactor safely.

6.20. Paragraph 6.34 of SSR-3 [1] states that "The challenges that the reactor may be expected to face during its operating lifetime shall be taken into consideration in the design process." Although it is not possible to apply a graded approach to this requirement, the design basis for items important to safety in a facility with a low potential hazard is typically less complex and requires less analysis to demonstrate that their performance meets acceptance criteria than in a facility with a high potential hazard. The classification of SSCs, based on their importance to safety (see paras 6.17 and 6.18), should be utilized to establish the design requirements for withstanding accident conditions without exceeding authorized limits.

Postulated initiating events

6.21. Requirements for identifying postulated initiating events for research reactors are established in Requirement 18 of SSR-3 [1].

6.22. The way this requirement to identify postulated initiating events is applied cannot be graded. A comprehensive set of postulated initiating events is always required for the safety analysis of a research reactor regardless of potential hazard and is required to be identified on the basis of engineering judgement, operating experience feedback (including operating experience from similar facilities) and deterministic assessment, complemented, where appropriate and available, by probabilistic methods (see para. 6.36 of SSR-3 [1]).

6.23. The analysis of the set of postulated initiating events should be commensurate with the potential hazard and the complexity of the research reactor. In particular, the scope and level of detail of the safety analysis should be commensurate with the characteristics of the design and the potential hazard of the research reactor (see paras 6.78–6.84).

Internal hazards and external hazards

6.24. Requirements for identifying and evaluating internal hazards and external hazards for research reactors are established in Requirement 19 of SSR-3 [1].

6.25. The identification of internal hazards (e.g. fire, explosion or flooding originating inside the research reactor) and external hazards (e.g. seismic activity, tornado or flooding external to the facility), that are applicable to the research reactor should be based on the site characterization and the design of the reactor. The application of this requirement cannot be graded. A detailed list of postulated internal and external hazards is included in appendix I to SSR-3 [1]. A graded approach can be used in applying the requirement to evaluate the effect of internal hazards and external hazards using safety analysis, based on the characteristics of the design and the potential hazard of the facility (see paras 6.78–6.84).

Design basis accidents

6.26. Requirements for identifying and considering design basis accidents for research reactors are established in Requirement 20 of SSR-3 [1].

6.27. The way this requirement to identify a set of design basis accidents based on postulated initiating events (see para. 6.21) is applied cannot be graded. Because the postulated initiating events will correspond to the degree of complexity and the potential hazard of the research reactor, the resulting design basis accidents will also reflect the facility design. For example, a critical assembly or subcritical assembly that does not need forced cooling flow might not have a design basis accident associated with loss of flow.

Design limits

6.28. Requirement 21 of SSR-3 [1] states:

> **"A set of design limits for a research reactor consistent with the key physical parameters for each item important to safety for the research reactor shall be specified for all operational states and for accident conditions."**

Design limits are limits on key physical parameters, such as the maximum stress or temperature that items are exposed to, that ensure the integrity of barriers and the reliability of safety functions. Design limits are also required to be specified for experimental devices (see para. 6.63 of SSR-3 [1]).

6.29. One aspect of Requirement 21 of SSR-3 [1] that can be applied using a graded approach is the degree of conservatism included in design limits. The specification of design limits should include conservatism to ensure that the limits are effective, are not exceeded and that the facility will withstand design basis accidents without acceptable limits for radiation protection being exceeded. The degree of conservatism can be adjusted in accordance with the potential hazard of the research reactor and the approach taken for safety analysis. For example, a research reactor with a low potential hazard could apply conservative design limits and simplify the safety analysis, whereas a research reactor with a higher potential hazard could apply less conservatism, leading to greater effort in a more detailed safety analysis.

Design extension conditions

6.30. Requirements for the derivation and use of design extension conditions for research reactors are established in Requirement 22 of SSR-3 [1]. The inclusion of design extension conditions in the safety analysis for a research reactor can use the overall graded approach for safety analysis described in paras 6.78–6.84 of this Safety Guide.

6.31. In applying a graded approach to the derivation of a set of design extension conditions, the potential hazard of the research reactor, engineering judgement and the results of the safety analysis for design basis accidents should be taken into account. In accordance with para. 6.64 of SSR-3 [1], the analysis of these design extension conditions could result in additional safety features. In a research reactor with a low potential hazard, such as some critical and subcritical assemblies with few SSCs important to safety, accidental criticality could be the only event included in the analysis of design extension conditions.

Engineered safety features

6.32. Requirements for engineered safety features for research reactors are established in Requirement 23 of SSR-3 [1].

6.33. For each design basis accident, the safety analysis for the research reactor should demonstrate that operational parameters are maintained within the specified design limits by the use of either passive or engineered safety features. As discussed in para. 6.30, the requirements for design limits may be applied using a graded approach, which in turn would have an effect on the design of engineered safety features.

6.34. The need for engineered safety features is identified by the safety analysis of the design of the research reactor. A research reactor with a high potential hazard and a large cooling system might need specific engineered safety features to mitigate internal flooding caused by a leak of secondary coolant. In addition, an emergency core cooling system (see Requirement 48 of SSR-3 [1]) might be required to collect and recirculate primary coolant inventory in response to a loss of coolant accident. For a research reactor with a low potential hazard, such as some critical assemblies where the irradiated fuel can be safely stored in air, the safety analysis may demonstrate that no engineered safety feature is necessary to maintain fuel integrity in response to a loss of coolant accident.

Reliability of items important to safety

6.35. Requirements for the reliability of items important to safety for research reactors are established in Requirement 24 of SSR-3 [1].

6.36. The use of a graded approach in the application of this requirement should be based on the potential hazard of the research reactor and the characteristics of the facility identified in the safety analysis. This analysis should demonstrate that the safety systems that prevent design limits from being exceeded (see para. 6.59 of SSR-3 [1]) operate with sufficient reliability. In accordance with a graded approach, the design of a safety system could use triplicate redundant channels to ensure high reliability; if even greater reliability is needed, the design could include a second system using diverse technology.

6.37. Where automatic or passive performance of a safety function is necessary or an inherent safety feature is used, a minimum level of reliability of the associated SSCs should be established and maintained. At least one automatic reactor shutdown system is required to be incorporated into the design (see para. 6.150 of SSR-3 [1]). Depending on the design basis of the research reactor, the performance of the following safety functions may also need to be automatic:

(a) Initiation of emergency core cooling;
(b) Confinement of radioactive material.

6.38. To ensure the necessary reliability of items important to safety at a research reactor one or more of the following design principles should be applied, as appropriate:

(a) Single failure criterion;
(b) Design for common cause failures;
(c) Physical separation and independence;
(d) Fail-safe design;
(e) Qualification of items important to safety.

Recommendations on the application of these principles to research reactors in accordance with a graded approach are provided in paras 6.39–6.46.

Single failure criterion

6.39. Requirement 25 of SSR-3 [1] states that "**The single failure criterion shall be applied to each safety group incorporated in the design of the research reactor.**"

6.40. A graded approach cannot be applied to the requirement that no single failure should prevent SSCs in a safety group from performing a main safety function. For all research reactors, the groups of equipment delivering any one of the main safety functions are required to be designed with appropriate redundancy, independence and diversity to ensure high reliability. However, the required degree of redundancy can be graded and may be lower for a facility with a low potential hazard.

Common cause failures

6.41. Requirement 26 of SSR-3 [1] states:

> "**The design of equipment for a research reactor facility shall take due account of the potential for common cause failures of items important to safety, to determine how the concepts of diversity, redundancy, physical separation and functional independence have to be applied to achieve the necessary reliability.**"

Because the objective is to achieve a level of reliability necessary to ensure safe operation, this requirement can be applied using a graded approach, for example, in the design of an emergency ventilation system. For a research reactor with a high potential hazard, where a design basis accident combined with the failure of emergency ventilation could result in off-site radiological consequences, to meet the acceptance criteria for the safety analysis, the design of the emergency ventilation system could exclude low-probability common cause failures through the use of diversity, redundancy and physical separation, whereas for a research reactor with a low potential hazard, the acceptance criteria may be met using a design based on simple redundancy of SSCs.

Physical separation and independence of safety systems

6.42. Requirements for the physical separation and independence of safety systems for research reactors are established in Requirement 27 of SSR-3 [1] and can be applied using a graded approach.

6.43. Physical separation can be incorporated into a design to varying degrees; for example, in a research reactor with a high potential hazard, system cable trains for two independent shutdown systems could be installed on separate floors of the facility to prevent a fault leading to a fire in one system affecting the second system. In a facility with a lower potential hazard, cable trains could be located in separate rooms or separated from each other within the same room and still meet the required reliability in the safety analysis for the system.

Fail-safe design

6.44. Requirement 28 of SSR-3 [1] states that "**The concept of fail-safe design shall be incorporated, as appropriate, into the design of systems and components important to safety for a research reactor.**"

6.45. In terms of applying a graded approach, engineering judgement should be applied, considering the acceptance criteria used in the safety analysis of the design of the research reactor, to assess the appropriate extent of fail-safe design features in systems and components important to safety, to ensure that safety functions are sufficiently reliable in response to initiating events to prevent and mitigate design basis accidents and selected design extension conditions.

Qualification of items important to safety

6.46. Requirement 29 of SSR-3 [1] states:

> "**A qualification programme shall be implemented for a research reactor facility to verify that items important to safety are capable of performing their intended functions when necessary, and in the**

prevailing environmental conditions, throughout their design life, with due account taken of reactor conditions during maintenance and testing."

In terms of applying a graded approach, the level of qualification of SSCs should be consistent with their safety classification (see paras 6.17 and 6.18).

Design for commissioning

6.47. Requirements for the design of a research reactor to facilitate commissioning are established in Requirement 30 of SSR-3 [1]. This requires that features to facilitate the commissioning process be included in the design "as necessary". Recommendations on the use of a graded approach in the application of requirements for commissioning of research reactors, including experimental devices and modifications are provided in SSG-80 [2].

Calibration, testing, maintenance, repair, replacement, inspection and monitoring of items important to safety

6.48. Requirements for the design to accommodate the calibration, testing, maintenance, repair, replacement, inspection and monitoring of items important to safety at a research reactor are established in Requirement 31 of SSR-3 [1].

6.49. The design of a research reactor is required to ensure that the performance of maintenance, periodic testing and inspection activities does not result in undue exposure to radiation of the operating personnel (see para. 6.88 of SSR-3 [1]). This aspect of the requirement is same irrespective of the potential hazard of the facility.

6.50. In designing a research reactor to facilitate the maintenance and testing of components during operation, the reliability of components (consistent with the manufacturer's recommendations and operating history) and their safety significance, as well as the potential hazard of the facility, should be taken into account. For example, for a research reactor with a high potential hazard, components in the reactor protection system might need testing more frequently than during shutdown periods. In such cases, the design should incorporate specific features to enable the testing of components or trains within a system without impairing the fulfilment of the safety function. In a facility with a lower potential hazard, the reliable performance of SSCs in the reactor protection system might be adequately demonstrated by testing performed during periodic shutdowns.

6.51. A graded approach can be applied to the storage and use of spare parts for maintenance of items important to safety, while still meeting regulatory requirements and applicable national codes and standards (e.g. admissible repair time) specified in the authorization and operational limits and conditions for the research reactor. For example, for a research reactor with a high potential hazard, spare parts for some SSCs important to safety might need to meet the national standards for nuclear power plants, including requirements for procurement and storage.

6.52. There are two steps in determining the provisions for maintenance, periodic testing and inspection of a research reactor, as follows:

(1) Firstly, the types and frequencies of inspections, tests and maintenance activities should be determined, taking into account the importance to safety of the SSC and its required reliability, and all of the effects that might cause progressive deterioration of the SSC.
(2) Secondly, the provisions to be included in the design to facilitate the performance of these inspections, tests and maintenance activities should be specified, with account taken of the frequency, the radiation protection implications and the complexity of the inspection, test or maintenance activity. These provisions may include accessibility, radiation shielding, remote handling and in situ inspection, self-testing circuits in electrical and electronic systems and software, and provisions for ease of decontamination and for non-destructive testing.

Design for emergency preparedness and response

6.53. Requirement 32 of SSR-3 [1] states:

"**For emergency preparedness and response purposes, the design for a research reactor facility shall provide:**

(a) **A sufficient number of escape routes, clearly and durably marked, with reliable emergency lighting, ventilation and other services essential to the safe use of these escape routes;**
(b) **Effective means of communication throughout the facility for use following all postulated initiating events and in accident conditions.**"

6.54. The way this requirement for escape routes to meet national requirements for emergency preparedness is applied cannot be graded. A graded approach can, however, be applied to the number, size and type of escape routes, which should be based on the layout and size of the facility, the number of personnel and the potential hazards in various zones of the research reactor. For a research reactor with a high potential hazard and a large number of operating personnel, the design of escape routes could be relatively versatile and the location where personnel assemble could need specific design features to protect personnel from hazards during an emergency. For a research reactor with a low potential hazard such as a critical assembly or a subcritical assembly with a small number of operating personnel, all the SSCs associated with the facility could be located in one or two rooms, and emergency routes could have correspondingly simpler designs.

6.55. A communication system for use in a research reactor with a high potential hazard and with several floors and rooms to accommodate the facility systems, a large number of operating personnel and elevated noise levels from equipment in some locations, could involve a complex design, including the ability to communicate via loudspeaker with specific rooms or zones within the building, and the ability for two-way communication between remote panels and the control rooms. The system design could also include diverse technology such as wired and wireless equipment, to increase its availability during an emergency. In a research reactor with a low potential hazard and a small number of operating personnel, where all the facility systems are contained in one or two rooms, a communication system could be of a simple design to allow control room personnel to provide warnings and instructions in an emergency.

Design for decommissioning

6.56. Requirements for design to support the decommissioning of a research reactor are established in Requirement 33 of SSR-3 [1]. This requirement also applies to design activities throughout the lifetime of the research reactor, including the design of modifications and new experimental devices and design activities in preparation for decommissioning.

6.57. A graded approach can be used in the selection of the design features for the protection of workers, the public and the environment, for example, as follows:

(a) Low potential hazard research reactors with small cores that are easily removed and packaged may need minimal special provisions for removal and packaging of the core. The need for disposal facilities for high level radioactive waste will likely be minimal.
(b) Pool type research reactors, which allow for underwater handling of the core components, may involve design provisions for disassembling the

reactor under the water. Facilities for the storage and disposal of radioactive waste will be an important consideration.

6.58. The provisions in the design to enable the decommissioning process should be based on the potential hazard of the research reactor; the power level, duration of operation and the associated levels of activation of core components; the predicted number and characteristics of other SSCs with radiological hazards (e.g. components in the primary coolant purification system); and the volume of material in the reactor building and reactor structure. In a research reactor with a low potential hazard, the used fuel and core components might need less additional shielding or specialized equipment for transport and storage than in a research reactor with a high potential hazard.

Design for radiation protection

6.59. Requirements for the design of a research reactor for radiation protection are established in Requirement 34 of SSR-3 [1]. This requirement can be applied using a graded approach, for example, in terms of the engineered features necessary to maintain doses below the established dose limits and as low as reasonably achievable, or the equipment to monitor and control access to the research reactor and its experimental devices.

6.60. Paragraph 6.94 of SSR-3 [1] requires that adequate provision is made for shielding, ventilation, filtration and decay systems in the design of a research reactor. For the design of ventilation systems, a graded approach can be used, based on the potential radiological hazard and the necessary occupancy of the room in operational states and in accident conditions. For a research reactor with a low or medium potential hazard, there are typically fewer locations within the facility requiring ventilation systems to mitigate radiological hazards than in a research reactor with a high potential hazard.

6.61. Design provisions to monitor and control access to SSCs imposing radiological hazards to workers can be applied using a graded approach. Based on the number of areas in the research reactor building with a radiological hazard that requires access control, the frequency of entry and the number of personnel, access control at a research reactor could be implemented using a range of design features from electronic locks with access cards, to a controlled set of keys administered by the control room, commensurate with the potential hazard of the research reactor and the complexity of the design.

6.62. Other requirements for radiation protection (see para. 6.4) and for radioactive waste management (see paras 6.14–6.16) at research reactors can also be applied using a graded approach and contribute to ensuring that doses do not exceed the established dose limits and are as low as reasonably achievable. Further recommendations on the use of a graded approach in the application of requirements for radiation protection and radioactive waste management for research reactors are provided in SSG-85 [7].

Design for optimal operator performance

6.63. Requirements for the design for optimal operator performance at a research reactor are established in Requirement 35 of SSR-3 [1]. These requirements can be applied using a graded approach to aspects of human factors and ergonomics, such as in the design of control room displays and audible signals for parameters important to safety and the development of operating procedures as a tool to prevent human errors.

6.64. The design of the human–machine interface in the control room can use a graded approach, based on the potential hazard of the research reactor, the number of SSCs important to safety and the corresponding number of parameters important to safety that need to be monitored. The depth of analysis of the human–machine interface can also use a graded approach. In all cases, the analysis of the human–machine interface should consider all normal operational states, postulated initiating events, design basis accidents and selected but enveloping design extension conditions, to ensure that combinations of alarms and indications in the control room are unambiguous.

6.65. For a research reactor with a low potential hazard, the number of SSCs is typically smaller with fewer operation and maintenance procedures than in a research reactor with a high potential hazard. The development of procedures involves expertise in human factors to assess the human–machine interface and the possible interactions between SSCs; this should also involve the application of a graded approach based on the size and complexity of the research reactor, the number of SSCs important to safety, and the potential consequences from an error made during operation or maintenance. Further recommendations on the development, use and improvement of operating procedures for research reactors are provided in SSG-83 [5].

Provision for safe utilization and modification

6.66. Design requirements for the safe utilization and modification of research reactors are established in Requirement 36 of SSR-3 [1]. The management system is required to include processes for new experiments and modifications to ensure a systematic approach to changes in the research reactor (see paras 4.16–4.19 of SSR-3 [1]). The main requirements for utilization and modification in the design of a research reactor are as follows:

(a) The reactor configuration is required to be known at all times (see para. 6.108 of SSR-3 [1]). The way this is applied cannot be graded. Configuration management is an important part of the design process (see para. 5.86 of IAEA Safety Standards No. GS-G-3.5, The Management System for Nuclear Installations [30]).

(b) New utilization and modification projects, including experiments that have a major or significant effect on safety, are required to be designed in accordance with the same principles that apply to the reactor (see para. 6.109 of SSR-3 [1]). This includes safety analysis (see also para. 6.78) and procedures for design construction, commissioning and decommissioning that are equivalent to those used for the research reactor itself. For less significant modifications and experiments, this element of the requirement can be applied using a graded approach, based on the potential hazard of the research reactor and the potential hazard of the proposed modification.

(c) Where experimental devices penetrate the reactor vessel or reactor core boundaries, they are required to be designed to preserve the means of confinement and reactor shielding (see para. 6.110 of SSR-3 [1]). The way this requirement is applied cannot be graded.

(d) Protection systems for experiments are required to be designed to protect the experiment and the reactor (see para. 6.110 of SSR-3 [1]). The way this requirement is applied cannot be graded.

6.67. SSG-24 (Rev. 1) [11] provides recommendations on designing and implementing new experiments or modifications at a research reactor. In particular, paras 3.7–3.12 of SSG-24 (Rev. 1) [11] provide recommendations on the use of a categorization process to determine the safety significance of an experiment or modification, and for the application of a graded approach based on this categorization. For a modification that is categorized as a 'major effect on safety' (see paras 3.13–3.20 of SSG-24 (Rev. 1) [11]), the operating organization should update the safety analysis for the research reactor and, as applicable, seek authorization from the regulatory body. The analysis of the modification should be reviewed by the reactor safety committee and the regulatory body. For a modification categorized as having a 'significant effect on safety' (see paras 3.21–3.28 of SSG-24 (Rev. 1) [11]), the existing safety analysis and authorization remain valid, but a change is needed in the operating limits and conditions for the research reactor. In such cases, an analysis should be performed to demonstrate the validity of the existing safety analysis report, and to justify the change in the operating limits and conditions. This analysis should be reviewed by the reactor safety committee and approved by the reactor manager before the design process proceeds. New or modified operating limits and conditions are required to be reviewed and approved by the regulatory body prior to commencement of operation with the modification or new experiment (see para. 7.33 of SSR-3 [1]). Recommendations on modifications categorized as having a 'minor effect on safety' or 'no effect on safety' are provided in paras 3.29–3.34 of SSG-24 (Rev. 1) [11].

6.68. Paragraph 9.6 of SSR-3 [1] includes the requirement for a change control process to evaluate new experiments or modifications and the effect the changes might have on safety and security. Technical guidelines on managing the interface between nuclear safety and security for research reactors are provided in Ref. [26].

6.69. The commissioning tests necessary to verify the acceptability of modifications is an aspect of this requirement that can be applied using a graded approach. For modifications and experiments with major safety significance, a formal commissioning programme is required (see para. 6.110 of SSR-3 [1]). Further recommendations on the use of a graded approach in the application of the requirement for a commissioning programme are given in paras 7.29–7.34 of this Safety Guide. Recommendations on the commissioning programme for modifications in research reactors are provided in SSG-24 (Rev. 1) [11] and SSG-80 [2].

Design for ageing management

6.70. Requirements for design to support ageing management for a research reactor are established in Requirement 37 of SSR-3 [1] and can be applied using a graded approach, based on the potential hazard, utilization and anticipated lifetime of the research reactor.

6.71. For a research reactor with a medium or low potential hazard, the ageing management programme, during the operation phase of the facility, should include a smaller number of items for monitoring, and fewer ageing management activities than the programme for a research reactor with a high potential hazard (i.e. which typically has more SSCs important to safety). A design with less-accessible SSCs might also be acceptable, providing the programme is able to verify the condition of all items important to safety and ensure that the necessary safety functions are fulfilled. A graded approach can be used in the application of this requirement, based on the safety classification of SSCs and expert judgement.

6.72. To take ageing management into account in the design of a research reactor, the use of materials resistant to degradation mechanisms, with sufficient design margins and provisions for testing, inspection and replacement, should be considered. The extent to which this is applied in the design can be determined using a graded approach, on the basis of the safety significance of the SSCs and their ease of replacement.

Provision for long shutdown periods

6.73. Requirements for long shutdown periods for research reactors are established in Requirement 38 of SSR-3 [1], and a graded approach can be used in the application of this requirement.

6.74. Research reactor designs normally include provisions to ensure safety during shutdown and typically these provisions can be used during a long shutdown. For all SSCs that are important to safety, and which could suffer degradation during a long shutdown period, provision should be made for a preservation programme that includes inspecting, testing, maintaining, dismounting and/or disassembling SSCs, as appropriate, during the shutdown period. As an alternative to implementing a preservation programme for installed equipment, it may be more practical to remove equipment; this decision is usually linked to the future of the research reactor. All modifications made to a research reactor in long shutdown are also subject to Requirements 36 and 83 of SSR-3 [1], including review, assessment and approval by the regulatory body prior to implementation, when appropriate.

6.75. The design of a fuel storage location for a long shutdown period can use a graded approach, based on the number of irradiated fuel assemblies, the total fission product inventory, the decay heat generated and the specific criticality and corrosion characteristics of the fuel assemblies. For a research reactor with a high potential hazard, the design could include a separate storage pool for irradiated fuel assemblies, equipped with heat removal and purification systems. Operating limits and conditions could be implemented (i.e. after review, assessment and approval by the regulatory body; see para. 7.33 of SSR-3 [1]) to prevent criticality safety events and to maintain the fuel assemblies in conditions where their integrity can be monitored and maintained. The design of the storage area, including cooling, purification and other support systems, should be based on safety analysis to ensure those systems are sufficiently reliable, applying the concept of redundancy and the single failure criterion. For a research reactor with a low potential hazard, such as some critical and subcritical assemblies with irradiated fuel containing a low fission product inventory that does not need shielding or water cooling, the irradiated fuel assemblies could be stored in a dry storage area of relatively simple design during a long shutdown period.

Prevention of unauthorized access to, or interference with, items important to safety

6.76. Requirement 39 of SSR-3 [1] states that "**Unauthorized access to, or interference with, items important to safety at a research reactor facility, including computer hardware and software, shall be prevented.**" The way this requirement is applied cannot be graded because preventing unauthorized access to nuclear facilities is necessary regardless of the size or potential hazard of the research reactor. Access controls are needed for operating personnel and other personnel involved in the operation or use of the reactor (e.g. technical support personnel and experimenters), as well as the public and emergency workers. A major objective of access control, in addition to preventing sabotage, is to prevent the unauthorized removal of nuclear material. Even research reactors with a low potential hazard and a low inventory of fission products in irradiated fuel assemblies, such as some critical and subcritical assemblies, should include specific design features for access control for those fuel assemblies to meet the objectives of safety and nuclear security.

Prevention of disruptive or adverse interactions between systems important to safety

6.77. Requirement 40 of SSR-3 [1] states:

"**The potential for disruptive or adverse interactions between systems important to safety at a research reactor facility that might be required to operate simultaneously shall be evaluated, and any disruptive or adverse interactions shall be prevented.**"

The way this requirement is applied cannot be graded because this evaluation is necessary for research reactors regardless of potential hazard. Design features to prevent disruptive or adverse interactions between systems are included in the safety analysis to demonstrate that systems important to safety perform reliably in response to all applicable initiating events. However, research reactors with a lower potential hazard typically have fewer systems important to safety, resulting in fewer adverse interactions between systems requiring evaluation.

Safety analysis of the design

6.78. Requirement 41 of SSR-3 [1] states:

> **"A safety analysis of the design for a research reactor facility shall be conducted in which methods of deterministic analysis and complementary probabilistic analysis as appropriate shall be applied to enable the challenges to safety in all facility states to be evaluated and assessed."**

6.79. The supporting requirements in paras 6.119–6.125 of SSR-3 [1] include several aspects that cannot be graded, for example, as follows:

(a) A safety analysis is required for all research reactors regardless of the potential hazard (see para. 6.119 of SSR-3 [1]).

(b) Use of the results from the safety analysis to define operational limits and conditions, form the design basis for items important to safety, and demonstrate adequate defence in depth in the design, is also required for all research reactors (see paras 6.119, 6.120 and 6.121(f) of SSR-3 [1]).

(c) Comparison of the results from the safety analysis with radiological acceptance criteria is required for all research reactors regardless of potential hazard (see para. 6.121(c) of SSR-3 [1]).

6.80. The safety analysis is also the basis for demonstrating the safety of the proposed design in support of an application for a licence and should be used to confirm that any use of a graded approach in the application of safety requirements has been appropriate.

6.81. The use of enveloping events in the safety analysis to include a range of input parameters, initial conditions, boundary conditions and assumptions can be applied using a graded approach. For a research reactor with a high potential hazard, the use of enveloping events, combining several such conditions, might not be possible if those enveloping events are too severe to meet the acceptance criteria. The safety analysis for such a research reactor typically makes limited use of enveloping events, and as a result includes a larger number of individual events for analysis. For a research reactor with a lower potential hazard, the conditions from separate events may be combined in enveloping events that, although more severe than any specific design basis accident, can be demonstrated to meet the acceptance criteria. The use of enveloping events for the safety analysis of these facilities simplifies the analysis process and needs fewer resources from the operating organization.

6.82. The scope and depth of the safety analysis should be based on the potential hazard of the facility, as discussed in para. 1.3 and annex I of Ref. [31]. The appendix to SSG-20 (Rev. 1) [10] provides recommendations on the content of the safety analysis report for research reactors and indicates where elements might not be applicable for subcritical assemblies. Paragraphs 3.1–3.7 of IAEA Safety Standards Series No. GSR Part 4 (Rev. 1), Safety Assessment for Facilities and Activities [15] establish requirements on a graded approach to safety assessment. Paragraph 3.3 of GSR Part 4 (Rev. 1) [15] states that "The main factor to be taken into consideration in the application of a graded approach is that the safety assessment shall be consistent with the magnitude of the possible radiation risks arising from the facility or activity."

6.83. The safety analysis for a facility with a relatively small number of SSCs and postulated initiating events would be simpler than that for a complex facility. Other examples of the application of a graded approach include the following:

(a)   The safety analysis may demonstrate that for some identified postulated initiating events the potential for a release of radioactive material from the core is physically impossible (or can be considered with a high level of confidence to be extremely unlikely), which would remove the need for extensive engineered safety features and analysis of their failure.
(b)   The presence of passive or inherent safety features and/or the absence of in-core experiments may also result in a reduction of the scope and depth of the safety analysis.
(c)   The use of conservative methods and criteria is a means of simplifying the safety analysis. Conservative criteria may be used in the safety analysis for research reactors with a low potential hazard.

6.84. A graded approach is required to be used in updating the safety assessment (see para. 3.7 of GSR Part 4 (Rev. 1) [15]). The frequency at which the safety assessment is updated and the level of detail of the safety assessment should be based on the following:

(a)   The number and extent of modifications to the research reactor systems and the safety significance of these modifications;
(b)   Changes to procedures;
(c)   Results of compliance monitoring of operational limits and conditions;
(d)   Evidence of component ageing;

(e) Results from research or internal and external operating experience;
(f) Changes in site conditions;
(g) Changes to input data used in safety analysis;
(h) New regulatory requirements.

THE USE OF A GRADED APPROACH IN SPECIFIC REQUIREMENTS
FOR THE DESIGN OF RESEARCH REACTORS

Buildings and structures

6.85. Requirements for buildings and structures for research reactors are established in Requirement 42 of SSR-3 [1].

6.86. A graded approach can be used for the design of shielding throughout the research reactor, based on the number of rooms where SSCs could be a source of radiation in operational states or in accident conditions, and on the characteristics of the radiation risk. In accordance with Requirement 42 of SSR-3 [1], the buildings and structures are required to be designed to maintain radiation levels as low as reasonably achievable and below the established dose limits. For a research reactor with a high potential hazard, a large number of rooms where equipment is associated with reactor operation, isotope production, experimental devices or radioactive waste storage could need to be provided with shielding as part of the building design. In a facility with a lower potential hazard, with a small number of rooms where a radiation risk is present, the design of structures to provide adequate shielding could be less complex.

6.87. Specific features in the design of buildings and structures will contribute to the application of a graded approach to other requirements of SSR-3 [1], for example, as follows:

(a) Separation of areas in accordance with their potential radiological hazard can minimize the need for radioactive waste handling, as well as contribute to the design for radiation protection, emergency preparedness and response, and fire protection.
(b) Up to date site evaluation can help to reduce excessive conservatism in the design of buildings and structures to ensure protection against external events (see section 2.2.1 of Ref. [31]).

Means of confinement

6.88. Requirements for the means of confinement for research reactors are established in Requirement 43 of SSR-3 [1]. The results of safety analysis, considering factors such as the fission product inventory in the core, and the proximity to population centres, can provide the basis for a graded approach in the application of this requirement.

6.89. For research reactors with a high potential hazard, safety analysis might demonstrate the need for a confinement system that includes a pressure-retaining containment structure (see footnote 25 and para. 6.137 of SSR-3 [1]) to meet the acceptance criteria. The necessary reliability of the safety functions performed by containment SSCs is determined by the acceptance criteria for off-site consequences under design basis accidents and selected design extension conditions. For a research reactor with a medium or low potential hazard, the reactor building could be designed without a pressure-retaining function, but with a ventilation system with features to control radioactive releases to meet the acceptance criteria. In all cases, the results of safety analysis should be used to determine how a graded approach is used in the design of the means of confinement, for example, whether iodine traps are necessary in the event of a release of fission products from the reactor.

Reactor core and fuel design

6.90. Requirements for reactor core and fuel design for research reactors are established in Requirement 44 of SSR-3 [1].

6.91. Paragraph 6.143 of SSR-3 [1] states that (footnote omitted) "The reactor core shall be designed so that the reactor can be shut down, cooled and maintained subcritical with an adequate margin for all operational states and accident conditions." The way this requirement is applied cannot be graded. However, a graded approach can, for example, be applied to the provisions in the design for monitoring the physical conditions and integrity of the fuel, and the analysis and experiments necessary to demonstrate the acceptability of the fuel.

6.92. For a research reactor with a high potential hazard, the monitoring of parameters such as temperature, flow and radiation levels in each fuel channel, could be a design feature that ensures an automatic response from the reactor protection system, or an action by operating personnel in response to an alarm. Such design features could be necessary to protect the research reactor in response to specific initiating events demonstrated in the safety analysis; however, the implementation of such a monitoring system could add additional SSCs to the research reactor design. In a facility with a lower potential hazard, bulk monitoring of coolant parameters such as pressure, temperature and radiation levels could be sufficient to demonstrate an adequate automatic response from safety systems and operator action in response to alarms following postulated initiating events.

6.93. The requirement in para. 6.138 of SSR-3 [1] to consider neutronic, thermohydraulic, mechanical, material, chemical and irradiation related factors in the design and qualification of fuel elements can be applied using a graded approach based on the potential hazard of the research reactor, and on existing analysis and qualification documents, including experience from other facilities. The extent of analyses and experiments necessary to demonstrate the acceptability of a reactor design with previously qualified fuel could be substantially smaller, particularly in a research reactor with a medium or low potential hazard, than that necessary for reactor designs that use new types of fuel assembly (i.e. where a fuel qualification process should be conducted).

Provision of reactivity control

6.94. Requirements for the provision of reactivity control for research reactors are established in Requirement 45 of SSR-3 [1]. Adequate reactivity control is required for all research reactor designs and the application of this requirement cannot be graded. Further recommendations on requirements for the main safety functions are provided in para. 6.3.

Reactor shutdown systems

6.95. Requirement 46 of SSR-3 [1] states:

> "**Means shall be provided for a research reactor to ensure that there is a capability to shut down the reactor in operational states and in accident conditions, and that the shutdown condition can be maintained for a long period of time, with margins, even for the most reactive conditions of the reactor core.**"

The way this requirement is applied cannot be graded.

6.96. Paragraph 6.152 of SSR-3 [1] states that "No single failure in the shutdown system shall be capable of preventing the system from fulfilling its safety function when required." This requirement is applied irrespective of the potential hazard of the facility.

6.97. Paragraph 6.155 of SSR-3 [1] states:

"It shall be demonstrated in the design that the reactor shutdown system will function properly under all operational states of the reactor and will maintain its reactor shutdown capability under accident conditions, including failures of the control system itself."

This requirement is applied irrespective of the potential hazard of the facility.

6.98. A graded approach can be used when determining how many redundant shutdown channels are necessary, how redundant channels will be credited in the safety analysis (see section 3 of Ref. [28]), and the extent of the instrumentation needed for monitoring the state of the shutdown system, based on the potential hazard of the research reactor.

Design of reactor coolant systems and related systems

6.99. Requirement 47 of SSR-3 [1] states that "**The coolant systems for a research reactor shall be designed and constructed to provide adequate cooling to the reactor core.**"

6.100.   The coolant system should be designed to provide adequate cooling to the reactor with an acceptable and demonstrated margin. Adequate cooling is required not only during normal operation at the authorized power levels, but also after shutdown (see para. 6.143 of SSR-3 [1]), under a range of anticipated operational occurrences and in accident conditions that involve loss of flow or loss of coolant transients. A graded approach can be used in the design of the coolant system. The coolant system can range from the provision of forced cooling with emergency electrical power being available to power some or all of the main coolant pumps, to no emergency power for any of the coolant pumps, to a system where natural convection cooling is used for both heat removal under full power operation as well as decay heat removal. Cooling by natural convection might be adequate for some low power research reactors.

6.101.   In a research reactor with a high potential hazard and a high power, the design of the SSCs to control the coolant temperature and pressure could be complex. In a research reactor with a medium potential hazard, SSCs to monitor water temperature, and pool volume could be of a simpler design while still meeting the requirements established in paras 6.73–6.81 of SSR-3 [1]. For a research reactor with a low potential hazard that does not have a heat removal system, such as some critical assemblies and subcritical assemblies, the safety analysis could confirm that there is no need to monitor certain parameters of the coolant such as pressure.

6.102.   The requirement to monitor and control the properties of the reactor coolant (e.g. the pH and conductivity; see para. 6.162 of SSR-3 [1]) is applicable to all water cooled research reactors of any power level, including some subcritical assemblies, to ensure that water conditions do not degrade reactor SSCs important to safety, especially boundaries that prevent the release of fission products, such as the fuel cladding.

Emergency cooling of the reactor core

6.103.   Requirement 48 of SSR-3 [1] states that "**An emergency core cooling system shall be provided for a research reactor, as required, to prevent damage to the fuel in the event of a loss of coolant accident.**" A graded approach can be used in the application of this requirement, based on the characteristics of the reactor and the fuel.

6.104.   The need for an emergency core cooling system should be defined in the design stage, and emergency operating procedures should be established, as necessary, taking into consideration the timescale needed for safe removal of the decay heat. To withstand a loss of coolant accident, a research reactor with a high potential hazard might need an emergency core cooling system to recover water lost from the primary cooling system, collect it in a sump and recirculate it back to cool the core. For a research reactor with a medium potential hazard, a simple system to replace the coolant inventory in the pool could be sufficient to prevent significant fuel failure due to a loss of coolant accident (see para. 6.164 of SSR-3 [1]). For a facility with a low potential hazard, such as some subcritical assemblies, the safety analysis could demonstrate that no emergency core cooling system is necessary to mitigate the consequences of a loss of coolant accident.

6.105.   For a research reactor where an emergency core cooling system is required, the system is required to perform its intended function in design basis accidents in the event of any single failure (see para. 6.165 of SSR-3 [1]). The way this requirement is applied cannot be graded.

THE USE OF A GRADED APPROACH IN INSTRUMENTATION AND CONTROL SYSTEMS FOR RESEARCH REACTORS

Provision of instrumentation and control systems

6.106.   Requirement 49 of SSR-3 [1] states:

> **"Instrumentation shall be provided for a research reactor facility for monitoring the values of all the main variables that can affect the performance of the main safety functions and the main process variables that are necessary for its safe and reliable operation, for determining the status of the facility under accident conditions and for making decisions for accident management."**

Most research reactors, irrespective of potential hazard, need, at a minimum, an emergency power supply for lighting (see Requirement 62 of SSR-3 [1]), instrumentation for monitoring the status of the facility (see Requirement 49 of SSR-3 [1]), emergency communication equipment (see Requirement 32 of SSR-3 [1]), and fire protection systems (see Requirement 61 of SSR-3 [1]), after a failure of normal electrical power.

6.107.   Aspects of the requirements for instrumentation and control systems can be applied using a graded approach based on the potential hazard of the facility, for example, in the provision of audio and visual alarms. In a research reactor with a high potential hazard there could be a large number of process variables and system parameters that necessitate audio or visual alarms or both to provide an early indication of changes in the operating conditions of the facility. Alarms may be necessary at locations other than the control room to ensure personnel are aware of the status of the research reactor and take appropriate action. In a research reactor with a low potential hazard such as some critical assemblies and subcritical assemblies, there could be a small number of process parameters that necessitate audio or visual alarms located in the control room. In all cases, the number of alarms and their location should be assessed in the safety analysis and in the emergency preparedness and response planning for the research reactor.

6.108.   In determining the types of measurement, locations of measurement and number of measurements to be taken of reactor parameters (e.g. temperature, pressure, flow, pool or tank water level, gamma radiation level, neutron flux and water chemistry parameters), the operational limits and conditions for the research reactor should be used as the basis for a graded approach. For example, the pressure drop across the core is measured in many reactors in order to detect reduced flow through the core. This measurement is typically not necessary in a research reactor that does not need active water cooling.

6.109.   A graded approach in the design of instrumentation and control systems can be based on the type of research reactor, the potential hazard and the role of the relevant SSCs, as stated in the safety analysis. Examples of design features that can be included, implemented in accordance with a graded approach, include the following:

(a)   Redundancy and diversity (see also para. 6.36);
(b)   Accuracy and precision;
(c)   Response time;
(d)   Level of quality assurance, as determined by the safety classification;
(e)   Level of automation.

6.110.   An example of a graded approach in the application of safety requirements for instrumentation and control systems is the choice of the level of redundancy. Triple channel redundancy is often used for research reactors that need to operate continuously, in order to minimize spurious scrams and to allow for testing and/or maintenance of instrumentation and control equipment during operation at power. For research reactors that operate for only a few hours per week or less frequently, such as some critical assemblies, a lower level (i.e. two channel (one-out-of-two)) of redundancy can be applied, thus reducing the complexity of the design and of operation.

Reactor protection system

6.111.   Requirement 50 of SSR-3 [1] states that "**A protection system shall be provided for a research reactor to initiate automatic actions to actuate the safety systems necessary for achieving and maintaining a safe state.**" The reactor protection system is required to automatically initiate the necessary protective actions for the full range of postulated initiating events to achieve a safe state (see para. 6.173 of SSR-3 [1]). A reactor protection system is required for all research reactor designs regardless of potential hazard.

6.112.   A graded approach can be applied to the reactor protection system, based on the potential hazard of the facility and the kind of initiating events identified in the safety analysis (based on consideration of, for example, potential consequences of the hazard, time constraints and mitigating passive safety features). For example, in a research reactor with a high potential hazard, there are typically a large number of SSCs important to safety and most of the postulated initiating events listed in appendix I to SSR-3 [1] are included in the design and the safety analysis. The reactor protection system in such a facility typically monitors a large number of process parameters to ensure that automatic action can be initiated in response to any postulated initiating event. In a research reactor with a lower potential hazard, natural convection cooling and no high pressure experimental devices, care should be taken in selecting postulated initiating events that are applicable for the reactor protection system design and safety analysis (e.g. primary pump failure, or loop rupture for a fuel testing experimental device). In such a research reactor, the reactor protection system could be designed with fewer sensors for process parameters, with correspondingly reduced complexity throughout the system. Other aspects of the research reactor will affect the design of the reactor protection system, including the following:

(a) At sites that could be affected by significant seismic events, a seismic sensor may be necessary to shut down the reactor, while at other sites with minimal seismic activity, such protection would not be necessary.
(b) Initiation of emergency core cooling may be necessary for certain reactors, while for others it would not be necessary (see paras 6.3(b)(iii) and 6.104–6.106).

6.113. Regardless of the hazard potential of the research reactor, the reactor protection system should be designed in such a way that neither a single failure nor a common cause failure will prevent meeting required safety functions.

Reliability and testability of instrumentation and control systems

6.114. Requirements for the reliability and testability of instrumentation and control systems for research reactors are established in Requirement 51 of SSR-3 [1] and can be applied using a graded approach.

6.115. In systems with high safety significance, such as a safety system in a research reactor with a high potential hazard, a design that includes a self-checking function within each channel of the instrumentation would allow an alarm to indicate a loss of function as soon as it occurred and minimize the time for which the fault was present. In systems important to safety where safety analysis has demonstrated that a loss of redundancy could exist for a defined period of time while the system continues to meet acceptable reliability targets, a function test should be performed at appropriate intervals (e.g. daily) to confirm the availability of each channel of instrumentation, and the design should support that level of testing. In a system with lower safety significance, the instrumentation and control equipment could be tested weekly or monthly and still perform with sufficient reliability.

Use of computer based equipment in systems important to safety

6.116.   Requirement 52 of SSR-3 [1] states:

> **"If a system important to safety at a research reactor is dependent upon computer based equipment, appropriate standards and practices for the development and testing of computer hardware and software shall be established and implemented throughout the lifetime of the system, and in particular throughout the software development cycle. The entire development shall be subject to an integrated management system."**

These requirements are applied irrespective of the potential hazard of the facility.

6.117.   Paragraph 6.184(g) of SSR-3 [1] states that "Appropriate verification and validation and testing of the software systems shall be performed." The way this requirement is applied cannot be graded.

Control room

6.118.   Requirements for the control room for research reactors are established in Requirement 53 of SSR-3 [1]. Based on the potential hazard of the research reactor and the accident conditions identified in the safety analysis report, the requirements for control room design can be applied using a graded approach. Under all conditions identified by safety analysis, the control room design is required to enable the research reactor to be maintained in (or returned to) a safe state. In a research reactor with a high potential hazard, accident conditions identified in the safety analysis could involve severe conditions due to a combination of radiation, hazardous chemicals, heat and humidity. In a research reactor with a low potential hazard, such as some critical assemblies and subcritical assemblies, the safety analysis might not identify any conditions that would necessitate additional protective measures in the control room. In all cases, the control room design should take into account the potential hazard of the research reactor and its environment, as well as the need for seismic resistance, ventilation systems and fire protection.

Supplementary control room

6.119.   Requirements for the supplementary control room for research reactors are established in Requirement 54 of SSR-3 [1]. The supplementary control room is required to support the fulfilment of the main safety functions, and the display of important parameters and radiological conditions in the facility (see para. 6.188 of SSR-3 [1]). A graded approach, based on the characteristics of the research reactor, the potential hazard and the accident conditions identified in the safety analysis report, can be used in the design of the supplementary control room or a remote shutdown panel. A graded approach could be used in determining the location, the number of parameters to be monitored and controlled and the actions necessary to maintain the reactor in a safe shutdown state. A graded approach could also be applied in relation to information from radiation monitors, fire detection systems and fire suppression systems in the research reactor, and emergency communication equipment.

6.120.   Paragraph 6.188 of SSR-3 [1] states that "A supplementary control room might not be necessary for critical assemblies and subcritical assemblies. In this case, the decision shall be justified on the basis of a comprehensive analysis." The safety analysis report for such a research reactor should demonstrate that the facility meets all acceptance criteria without a supplementary control room being included in the design.

Emergency response facilities on the site

6.121.   Requirements for emergency response facilities on the site of a research reactor are established in Requirement 55 of SSR-3 [1]. The requirements for the scope and functions of emergency response facilities can be applied using a graded approach, based on the nature and severity of the accident conditions identified in the safety analysis report, together with other emergency scenarios included in the scope of the design for the emergency response facilities. Aspects of this requirement that could be applied using a graded approach include the structure and number of the emergency response facilities and the provision of information and equipment for communication.

6.122.   For a research reactor with a high potential hazard, the conditions near the on-site emergency response facilities during an emergency could be hazardous, including high radiation levels. To respond adequately to emergencies, separate emergency response facilities could be necessary to protect personnel involved in emergency response and to support, in an integrated manner, the provision of technical support, operational support and on-site emergency management. For a research reactor with a low potential hazard, such as some subcritical assemblies, where the safety analysis does not identify a significant hazard outside the reactor building as a result of any design basis accident, the emergency response facility could be of a simpler design, with no special protective measures.

Electrical power supply systems

6.123.   Requirements for electrical power supply systems for research reactors are established in Requirement 56 of SSR-3 [1]. These requirements can be applied using a graded approach based on a variety of factors, including the potential hazard of the research reactor, the type and number of safety functions and engineered safety features for which normal or emergency power is needed, and the accident conditions that the electrical power supply needs to withstand. The reliability requirements might be different for different reactors, for the various utilization programmes of a particular reactor and for the needs of experimental devices. In applying a graded approach, the number, size and reliability of any necessary emergency power supply systems should be considered.

6.124.   For a research reactor with a high potential hazard, where forced cooling is needed to remove decay heat, the level of redundancy and the number of separate channels in the emergency power supply system should be based on the results of safety analysis. The duration for which the emergency power supply needs to deliver power should be based on the characteristics of the fuel and the nature of the accident conditions. In a research reactor with a low potential hazard, such as some critical assemblies and subcritical assemblies with very low inventories of fission products and no significant decay heat, emergency power for cooling is not necessary.

6.125.   A graded approach can be used in the application of the requirement for emergency power for the communication system (see para. 6.55). The emergency power supply to the communication system should be of commensurate design and reliability.

Radiation protection systems

6.126.   Requirements for radiation protection systems for research reactors are established in Requirement 57 of SSR-3 [1]. The requirements for radiation protection systems can be applied using a graded approach, to ensure that the design of radiation protection systems provides adequate monitoring for the research reactor and is commensurate with the nature and extent of the radiological hazards. Paragraph 6.193 of SSR-3 [1] lists the radiation protection systems used in research reactor facilities and the purposes they serve. Each of these systems should be considered in the design for a research reactor, regardless of potential hazard.

6.127. Examples of considerations in the application of a graded approach to radiation monitoring include the following:

(a) The number and extent of deployment of fixed radiation monitoring instruments should be commensurate with the potential hazard of the research reactor and the number of rooms or areas where a potential radiological hazard could arise during operational states and accident conditions.
(b) A research reactor with SSCs in areas where a radiological hazard from neutrons may be present (e.g. beam tubes and neutron guides) should have sufficient neutron and gamma monitors near those SSCs, as well as equipment for monitoring contamination.
(c) A research reactor with a low potential hazard (e.g. used for teaching purposes) might need only limited monitoring equipment, such as gamma radiation monitors at the open pool end or in the control console, and contamination monitors.
(d) For research reactors with a high potential hazard and a large number of personnel, supplementary monitoring displays elsewhere in the facility (i.e. outside the control room) should be provided for displaying the radiological conditions at specific locations in the facility in operational states and in accident conditions (wide range monitoring).

Handling and storage systems for fuel and core components

6.128. Requirements for the handling and storage systems for fuel and core components for research reactors are established in Requirement 58 of SSR-3 [1]. The aim of this requirement is to ensure safety in the handling and storage of fresh and irradiated fuel, core components and experimental devices. The main concerns are the prevention of inadvertent criticality and fuel damage from mechanical impacts, corrosion or other chemical damage. Two elements of these requirements are applied irrespective of the potential hazard of the facility, namely:

(a) The requirement to prevent criticality by an adequate margin (see para. 6.198(a) of SSR-3 [1]);
(b) The requirement to enable individual fuel elements and assemblies to be identified and tracked (see para. 6.198(i) of SSR-3 [1]).

6.129. The application of other elements of the requirements can use a graded approach, based on the potential hazard of the research reactor, the design of the reactor and its utilization programme. For example, the design of the storage location for irradiated fuel could be a separate fuel storage pool with systems for cooling and purification, or an area within the reactor pool designated for fuel storage, or, for a research reactor with a low potential hazard (e.g. some critical assemblies and subcritical assemblies), the irradiated fuel assemblies could be safely stored in a dry storage area in the reactor hall.

6.130. A graded approach for the design of storage systems should be based on the storage needs for all types of irradiated fuel assembly used in the research reactor, and for experimental fuel as well as experimental devices or equipment and materials used in isotope production. Other considerations include the means of decay heat removal and protection from mechanical impacts or corrosion.

Radioactive waste systems

6.131. General requirements for the predisposal management of radioactive waste are established in IAEA Safety Standards Series No. GSR Part 5, Predisposal Management of Radioactive Waste [32]. Requirements for the design of radioactive waste systems at research reactors are established in Requirement 59 of SSR-3 [1]. The operating organization should apply a graded approach in the application of the requirements for the management of radioactive waste, and for control and monitoring of solid, liquid and gaseous discharges, based on the types and quantities of radioactive waste generated in the research reactor. The operating organization should apply a graded approach to the design of shielding in radioactive waste systems, based on the characteristics and radiological hazard of the waste.

6.132. IAEA Safety Standards Series No. SSR-6 (Rev. 1), Regulations for the Safe Transport of Radioactive Material, 2018 Edition [33], includes a graded approach to performance standards for package designs for the safe transport of radioactive material, including radioactive waste, and the appendix to IAEA Safety Standards Series No. TS-G-1.4, The Management System for the Safe Transport of Radioactive Material [34], provides detailed examples of a graded approach for all aspects of the transport of radioactive material.

THE USE OF A GRADED APPROACH IN SUPPORTING SYSTEMS AND AUXILIARY SYSTEMS FOR RESEARCH REACTORS

Performance of supporting systems and auxiliary systems

6.133.   Requirements for the performance of supporting systems and auxiliary systems for research reactors are established in Requirement 60 of SSR-3 [1]. A research reactor with a lower potential hazard typically has fewer and simpler SSCs important to safety, including supporting and auxiliary systems. In accordance with Requirement 60 of SSR-3 [1], the design of supporting systems and auxiliary systems is required to be commensurate with those systems that they support; consequently, the way this requirement is applied cannot be graded as each system needs to comply with the design and the performance characteristics stated in the safety analysis.

Fire protection systems

6.134.   Requirements for fire protection systems for research reactors are established in Requirement 61 of SSR-3 [1]. Requirements for fire protection systems can be applied using a graded approach based on the results of safety analysis, fire hazard analysis and/or expert judgement, provided that the systems remain in compliance with regulatory requirements. For example, fire protection systems are required to provide alarms and information on the location of fires (see para. 6.207 of SSR-3 [1]). In a research reactor with a high potential hazard, there are typically a large number of rooms on different floors of the reactor building, whereas a research reactor with a low potential hazard could be located in a single reactor hall. In addition, based on the results of a fire hazard analysis and the layout of the facility, the information displayed by the fire protection system could vary in scope and complexity.

Lighting systems

6.135.   Requirements for lighting systems for research reactors are established in Requirement 62 of SSR-3 [1]. These requirements can be applied using a graded approach on the basis of safety analysis and expert judgement. Safety analysis should identify where actions by operating personnel are necessary in response to accident conditions, and which areas of the reactor building would need to be accessed during an emergency response. The outcome of this analysis should be used as the basis for the design of lighting systems. For a research reactor with a high potential hazard, lighting and emergency lighting systems could be extensive and include an emergency electrical power supply. For some research reactors with a low potential hazard, the facility may be located in a single reactor hall where the provision of adequate lighting is straightforward.

Lifting equipment

6.136.   Requirements for lifting equipment for research reactors are established in Requirement 63 of SSR-3 [1].

6.137.   The design of lifting equipment in a research reactor is required to prevent the lifting of excessive loads, prevent the dropping of loads that could cause a radiological hazard, permit the safe movement of lifting equipment, be seismically qualified if used in areas where equipment important to safety is located and permit periodic inspection (see para. 6.210 of SSR-3 [1]). The way these requirements are applied cannot be graded.

Air-conditioning systems and ventilation systems

6.138.   Requirements for air-conditioning systems and ventilation systems for research reactors are established in Requirement 64 of SSR-3 [1] and can be applied using a graded approach. For a research reactor with a high potential hazard, the design may include normal and emergency ventilation systems based on the results of safety analysis and the characteristics and locations of potential airborne radiological hazards. If the research reactor has a potential tritium hazard, the ventilation system may include additional features to detect and mitigate that hazard. For a research reactor with a low potential hazard, based on the results of safety analysis, monitoring for airborne radioactivity could be performed by periodic checks on an air filter, with no other special ventilation equipment needed in the design.

Compressed air systems

6.139.   Requirements for compressed air systems for research reactors are established in Requirement 65 of SSR-3 [1]. For a compressed air system serving an item important to safety at a research reactor, the design is required to specify three parameters: quality, flow rate and cleanness; the way this requirement is applied cannot be graded.

Experimental devices

6.140.   Requirement 66 of SSR-3 [1] states:

> **"Experimental devices for a research reactor shall be designed so that they will not adversely affect the safety of the reactor in any operational states or accident conditions. In particular, experimental devices shall be designed so that neither the operation nor the failure of an experimental device will result in an unacceptable change in reactivity for the reactor, affect operation of the reactor protection system, reduce the cooling capacity, compromise confinement or lead to unacceptable radiological consequences."**

The requirement that the operation or failure of an experimental device does not result in the consequences listed above is applied irrespective of the potential hazard of the facility; these consequences are required to be prevented in all research reactors.

6.141.   A graded approach can be applied to some aspects of the design of experimental devices, based on the potential hazard of both the research reactor and the experimental device. For example, a graded approach can be applied to the design of alarm signals and trip signals of experiments interconnected with the reactor protection system, and/or the control signals of the experiment connected to the reactor instrumentation and control system. For a research reactor with a high potential hazard and involving an experimental device that affects the reactivity of the core, such as a fuel testing facility, the experimental device could include specific instrumentation for the reactor protection system to initiate a scram. In the same reactor, a simple experimental device for performing irradiations that do not need cooling might not include any instrumentation in its design. A graded approach can also be applied to the monitoring of the experimental devices from the control room(s).

6.142.   The design, analysis and the authorization process for experimental devices (see also paras 7.70–7.75) should be commensurate with the potential hazard of both the research reactor and the experimental device, with the operating organization's familiarity with the experiment, and with any existing, relevant safety analyses. For the installation of a new experimental device where the potential hazard is high, and a failure of the experimental device represents a new initiating event outside the scope of the safety analysis report, a revision of the safety analysis report is required (see para. 6.212 of SSR-3 [1]), and any necessary revision of the operational limits and conditions is required to be submitted to the regulatory body for review, assessment and approval prior to commencement of operation with the new experimental device (see para. 7.73 of SSR-3 [1]). For an experimental device with a low potential hazard, and which is equivalent to experiments that have previously been installed in the research reactor (e.g. an irradiation experiment that does not need active cooling), the safety analysis and regulatory approval could be simplified by confirming that the irradiation conditions are bounded by those in the existing safety analysis. Recommendations on a categorization process for experimental devices are provided in section 3 of SSG-24 (Rev. 1) [11].

7. THE USE OF A GRADED APPROACH IN THE OPERATION OF RESEARCH REACTORS

THE USE OF A GRADED APPROACH IN ORGANIZATIONAL PROVISIONS FOR A RESEARCH REACTOR

Responsibilities of the operating organization

7.1. Requirements on the responsibilities of the operating organization of a research reactor are established in Requirement 67 of SSR-3 [1]. Recommendations on meeting these requirements are provided in SSG-84 [6].

7.2. The general responsibilities and functions of the operating organization as well as responsibilities, functions and lines of communication of the key positions within the reactor operation organization (see paras 7.1–7.7 of SSR-3 [1]), apply equally to all research reactors regardless of their potential hazard. Similarly, a graded approach cannot be applied to staff positions that require a licence or certificate in accordance with the legal framework of the State (see para. 7.5 of SSR-3 [1]).

7.3. For all research reactors, the direct responsibility and the necessary authority for the safe operation of the reactor are required to be assigned to the reactor manager and cannot be delegated (see para. 7.3 and Requirement 69 of SSR-3 [1]).

7.4. Paragraph 7.2 of SSR-3 [1] states:

"The operating organization shall ensure that adequate provision is made for all functions relating to the safe operation and utilization of the research reactor facility, such as maintenance, periodic testing and inspection, radiation protection, quality assurance and relevant support services."

The manner in which these functions can be performed could be subjected to the use of a graded approach in accordance with their safety significance, and the maturity and complexity of the research reactor. For example, in a research reactor with a low potential hazard and which is subject to effective quality checks, the maintenance, periodic testing and inspection activities could be performed by the reactor operators.

7.5. The implementation of a management system (see para. 7.1 of SSR-3 [1]) is an aspect of this requirement that can be applied using a graded approach (see also paras 4.6–4.11 of this Safety Guide). For example, a research reactor with a high potential hazard will involve a large amount of management system documentation on roles and responsibilities, processes for operation and maintenance of reactor SSCs, and programmes for radiation protection, ageing management, environmental monitoring, waste management and utilization. In comparison, the management system for a research reactor with a low potential hazard is likely to need less documentation.

Structure and functions of the operating organization

7.6. Requirements for the structure and functions of the operating organization of a research reactor are established in Requirement 68 of SSR-3 [1]. The requirements in para. 7.11 of SSR-3 [1] for the organizational structure to be documented (including the roles that are critical for safe operation), and for changes to the documented organizational structure to be analysed before implementation, is applied irrespective of the potential hazard of the facility.

7.7. The use of a graded approach in the application of other aspects of the requirements for the structure and functions of the operating organization should be based on the potential hazard of the research reactor and the national legal and regulatory framework. For example, a research reactor in a State with a limited nuclear programme might need a large and complete in-house capability (e.g. a technical support group, expertise in quality control, a large inventory of spare components, expertise in isotope production and maintenance personnel). In comparison, a similar research reactor in a State with a large infrastructure and nuclear programme might not need such a large in-house capability because support could be easily obtained from external organizations.

7.8. The use of a graded approach in the application of requirements on the structure and functions of the operating organization can be applied in the following areas:

(a) The number and duties of operating personnel. For a research reactor with a low potential hazard (i.e. which is typically less complex and has fewer SSCs compared with a research reactor with a medium or high potential hazard), an individual could be assigned multiple duties. In such cases, arrangements should be established to ensure functional independence (e.g. in the radiation protection function) and effective quality checks.
(b) Membership and frequency of meetings of the reactor safety committee (see paras 4.14 and 7.10).
(c) Preparation and periodic updating of the safety analysis report (see SSG-20 (Rev. 1) [10]).
(d) Training, retraining and qualification programmes (see SSG-84 [6] and paras 7.11–7.15).
(e) Operating procedures (see SSG-83 [5] and paras 7.35–7.39).
(f) Maintenance, periodic testing and inspection programmes (see SSG-81 [3] and paras 7.43–7.52).
(g) Emergency planning and procedures (see paras 7.63–7.67).
(h) The radiation protection programme (see SSG-85 [7] and paras 7.75–7.81).
(i) The management system (see Section 4).

Operating personnel

7.9. Requirements for operating personnel for a research reactor are established in Requirement 69 of SSR-3 [1]. Irrespective of the potential hazard of the research reactor, the key positions within the operating organization include the reactor manager, operating personnel, including maintenance staff, radiation protection personnel, additional support staff such as training officers and safety officers, and reactor safety committee members. However, the number of personnel in some of these positions should be subjected to the use of a graded approach. For example, a larger number of operating personnel are typically needed for a research reactor with a high potential hazard, depending on its operating schedule (e.g. operation in shifts), and other factors, such as the level of automation and the number of maintenance activities. See para. 7.8(a) for recommendations on the use of a graded approach in the application of this requirement on the number of operating personnel.

7.10. A reactor safety committee is required for all research reactors, with the responsibilities described in paras 7.26–7.27 of SSR-3 [1]. A graded approach should be used in the application of this requirement with respect to the number of members of the safety committee, including their appropriate level and range of technical expertise and the frequency of meetings, based on the potential hazard and the utilization schedule of the research reactor, and the number and complexity of planned modifications with safety significance.

Training, retraining and qualification of personnel

7.11. Requirements for training, retraining and qualification of personnel for a research reactor are established in Requirement 70 of SSR-3 [1]. Recommendations on meeting these requirements are provided in SSG-84 [6].

7.12. In accordance with Requirement 70 of SSR-3 [1], a training programme for the training, retraining and qualification of research reactor personnel is required regardless of the potential hazard of the facility. The need for a systematic approach to training, including assessment of training needs, and the design, development, implementation and evaluation of both initial and continuing training is applicable to all research reactors. However, a graded approach should be used in the training programme, which should be consistent with the complexity of the research reactor design, the potential hazard, the planned utilization of the research reactor and the functions that might be assigned to the personnel being trained.

7.13. A graded approach could also be applied to the education level and operating experience of trainees, the content and duration of initial and continuing training, training materials, the assessment of completed training, and qualification, which can depend on the complexity of the research reactor design, as well as the potential hazard, planned utilization and available infrastructure.

7.14. The training programme should cover theoretical and facility-specific knowledge, as recommended in SSG-84 [6]. The contents and duration of the theoretical training should be the same for all research reactors; however, training on facility-specific knowledge should be more extensive for research reactors with a high potential hazard and/or with more complex designs. The topics included in a continuing training course and the appropriate duration are also dependent on any recent changes to SSCs, operating procedures and operational limits and conditions. The duration of continuing training could be a few days per year for research reactors with a low or medium potential hazard. For complex research reactors with a high potential hazard, the duration of continuing training could be up to a few weeks per year. See also section 4 of and annex II to SSG-84 [6] for recommendations and guidance on the contents and duration of initial and continuing training of operating personnel for research reactors.

7.15. A graded approach to the application of the requirement for reauthorization (see para. 7.28 of SSR-3 [1]) after absences (see para. 5.13 of SSG-84 [6]), should ensure that retraining, requalification and examinations are commensurate with the duration of the absence, the complexity and potential hazard of the research reactor, and the changes to the facility and its operation during the absence of the individual. For example, in a research reactor with a high potential hazard, a significant amount of retraining for a reactor operator could be necessary, whereas for a research reactor with a lower potential hazard, retraining after a similar absence could be accomplished in less time.

THE USE OF A GRADED APPROACH IN OPERATIONAL LIMITS AND CONDITIONS FOR A RESEARCH REACTOR

7.16. Requirements for operational limits and conditions for a research reactor are established in Requirement 71 of SSR-3 [1]. Recommendations on the preparation and implementation of operational limits and conditions for research reactors are provided in SSG-83 [5].

7.17. Operational limits and conditions are based on the reactor design and on the information from the safety analysis report; consequently, a graded approach should be used in the application of the requirements for design and safety analysis, as discussed in Sections 3 and 6 .

Safety limits

7.18. Safety limits should be established in the design stage for a research reactor, based on the results of the safety analysis. Paragraph 7.35 of SSR-3 [1] states that "Safety limits shall be set to protect the integrity of the physical barriers that protect against the uncontrolled release of radioactive material or exposure over regulatory limits." The way this requirement is applied cannot be graded. For example, the value of the safety limit on the maximum cladding temperature should be based on the physical properties of the cladding material and its environment, regardless of the potential hazard of the research reactor. However, the depth of analysis that is used to establish the safety limit should vary depending on the potential hazard of the research reactor.

Safety system settings

7.19. Paragraph 7.36 of SSR-3 [1] states that "Safety system settings shall be defined so that the safety limits are not exceeded."

7.20. For each safety limit, at least one safety system is needed to monitor parameters and to provide a signal to accomplish an action (e.g. to shut down the reactor) in order to prevent the parameter from approaching the safety limit. The safety system setting should be at an acceptable safety margin from the safety limit. For protective safety actions of particular importance, such as neutronic trips (scrams), redundant systems should be employed. The depth of the analysis performed, including the use of methods to evaluate uncertainty, to establish a safety margin can be based on a graded approach. The minimum value of an acceptable safety margin could be determined using of a graded approach that is commensurate with the potential hazard of the research reactor.

7.21. A graded approach might also be applied in relation to the redundancy and diversity of instruments, in particular to the selection of the types of safety system setting relating to the safety limits and the operational limits and conditions. For example, in a low power reactor, the coolant outlet temperature could be selected as the parameter relating to the fuel temperature for which a safety system setting is defined, while in a higher power reactor, to prevent the safety limits from being approached, a complex system of variables should have defined safety system settings, such as the coolant outlet temperature, the inlet temperature, the coolant flow rate, the differential pressure across the core and the primary pump discharge pressure, as well as parameters for experimental facilities. In addition, two safety parameters (e.g. pressure and flow) may also be needed for detection of some design basis accidents.

Limiting conditions for safe operation

7.22. Limiting conditions for safe operation are operational constraints and administrative limitations on parameters and equipment that are established to provide acceptable margins between normal operating values and the safety system settings during startup, operation, shutting down and shutdown of a research reactor. The selection of the number and scope of limiting conditions for safe operation of a research reactor is an aspect of this requirement that can be applied using a graded approach. For example, a research reactor with a high potential hazard typically has more SSCs important to safety and a greater number of parameters for which limiting conditions for safe operation need to be specified than a research reactor with a medium or low potential hazard and fewer SSCs important to safety. Appendix I to SSG-83 [5] provides a list of operating parameters and equipment to be considered in establishing limiting conditions for safe operation. A graded approach could be used to determine the type and depth of analysis performed in establishing a limiting condition for safe operation in accordance with the type of reactor and conditions of operation.

Requirements for maintenance, periodic testing and inspection

7.23. As part of the operational limits and conditions for a research reactor, the operating organization is required to establish requirements for maintenance, periodic testing and inspection of items important to safety (see paras 7.38 and 7.39 of SSR-3 [1]). In order to ensure that safety limits and limiting conditions for safe operation are met, the relevant SSCs are required to be reliable and available. To ensure adequate reliability, SSCs important to safety are required to be maintained, monitored, inspected, checked, calibrated and tested in accordance with approved maintenance, periodic testing and inspection programmes (see Requirement 77 of SSR-3 [1]). Surveillance requirements in the operational limits and conditions specify the frequency and scope of inspections and acceptance criteria for each SSC. A graded approach should be used in the application of these requirements on the basis of the importance to safety and the required reliability of each SSC. Additional recommendations are provided in paras 7.43–7.52.

Administrative requirements

7.24. Administrative requirements include those for the organizational structure and responsibilities, minimum staffing, training and retraining, review and audit procedures, records and reports, and event investigation and follow-up (see para. 7.40 of SSR-3 [1]). In a research reactor with a high potential hazard and continuous operation day and night, the operational limits and conditions could specify several administrative requirements for shift turnover, minimum staffing levels, technical specialists such as chemistry or radiation protection personnel, operating logs and reporting of events; these administrative requirements might not all be needed (or at least not needed in the same level of detail) for research reactors with a medium or low potential hazard or those that have a limited operation schedule.

Violations of operational limits and conditions

7.25. The requirements for actions after a violation of operational limits and conditions in paras 7.41–7.43 of SSR-3 [1] are applied irrespective of the potential hazard of the facility. The nature of the action will be determined by the regulatory framework of the State and will typically depend on the severity of the violation.

THE USE OF A GRADED APPROACH IN THE PERFORMANCE OF SAFETY RELATED ACTIVITIES AT A RESEARCH REACTOR

7.26. Requirements for the performance of safety related activities at a research reactor are established in Requirement 72 of SSR-3 [1].

7.27. Paragraph 7.44 of SSR-3 [1] states:

"All routine and non-routine operational activities shall be assessed for potential risks associated with harmful effects of ionizing radiation. The level of assessment and control shall depend on the safety significance of the task."

7.28. For a research reactor with a high potential hazard, the operating organization could include a group of personnel to plan, assess and control operation and maintenance tasks. For research reactors with a lower potential hazard, a smaller number of operation and maintenance tasks could be planned, assessed and controlled by the same personnel who operate and maintain the facility. Expertise in radiation protection is necessary for all research reactors to assess tasks involving exposure to radiation.

THE USE OF A GRADED APPROACH IN COMMISSIONING OF A RESEACH REACTOR

7.29. Requirements for the commissioning programme for a research reactor are established in Requirement 73 of SSR-3 [1]. Recommendations on meeting Requirement 73 are provided in SSG-80 [2].

7.30. A commissioning process is required for all SSCs, activities and experiments regardless of the potential hazard of the research reactor. However, a graded approach can be used in the application of the requirement for a commissioning programme in the following areas:

(a) Organization for commissioning;
(b) Commissioning tests and stages;
(c) Commissioning procedures and reports.

7.31. In accordance with Requirement 73 of SSR-3 [1], an organizational structure for commissioning, including for utilization and modifications important to safety, is required regardless of the potential hazard of the research reactor. However, the number of personnel within this structure (including the number of personnel from the operations group) and the necessary expertise can vary depending on the potential hazards of the research reactor and its design. For example, for some research reactors, critical assemblies and subcritical assemblies with a low potential hazard, the organizational structure for commissioning typically includes fewer personnel from the operations group and less (or even no) expertise on power rise tests and operation at high power levels.

7.32. Stage C of commissioning (power ascension tests and power tests up to rated full power as defined in para. 3.17 and paras 5.30–5.37 of SSG-80 [2]) is not necessary for some critical and subcritical assemblies with a low potential hazard, and the scope, extent, and duration of Stage C are much less for low power research reactors (i.e. typically with a low potential hazard) compared with those with higher power levels. The scope, number and types of commissioning tests, procedures and reports, as well as the number of hold points in the commissioning process are much smaller for research reactors with a low potential hazard and less complex design, compared with those for facilities with a medium or high potential hazard. The number of hold points in the commissioning process should be determined taking into account the potential hazard of the research reactor and the safety significance of the subsequent step in the commissioning procedure. Regardless of the potential hazard of the research reactor, there should always be a hold point for tests prior to fuel loading (pre-operational tests).

7.33. A graded approach to commissioning tests for a research reactor should be adopted, as described in para. A.2 of SSG-80 [2]). The extent and type of tests to be performed should be determined on the basis of the importance to safety of each item and the potential hazard of the research reactor. Further recommendations on the use of a graded approach in application of safety requirements for the commissioning of research reactors are provided in SSG-80 [2].

7.34. The principles applied in commissioning for the initial approach to criticality, reactivity device calibrations, neutron flux measurements, determination of core excess reactivity and shutdown margins, power raising tests and testing of the containment system or other means of confinement are similar for all research reactors regardless of potential hazard and hence cannot be subject to a graded approach.

THE USE OF A GRADED APPROACH IN THE OPERATION OF A RESEARCH REACTOR

Operating procedures

7.35. Requirements for operating procedures for research reactors are established in Requirement 74 of SSR-3 [1]. Recommendations on the preparation of operating procedures are provided in SSG-83 [5]. Appendix II to SSG-83 [5] presents an indicative list of operating procedures for research reactors.

7.36. Prior to operation, a graded approach should have been used in the application of the requirements for research reactor design, construction, commissioning and safety analysis, including the development of operational limits and conditions, based on the potential hazard, the design and the complexity of the facility. In addition, a graded approach should have been used in the application of the requirements for the establishment and implementation of the management system that governs the format, development, review, control and training on the use of operating procedures for the research reactor. The use of a graded approach in the application of safety requirements for operating procedures should be consistent with the use of that approach in these programmes and activities.

7.37. The list of operating procedures presented in appendix II to SSG-83 [5] should be assessed for applicability to a specific research reactor. The number of operating procedures developed should be dependent upon the characteristics of the research reactor and should be less for reactors with fewer SSCs important to safety and a low potential hazard. For example, in a facility with a low potential hazard, such as some critical assemblies and subcritical assemblies, procedures related to the surveillance of systems such as cooling and ventilation systems might not be necessary, and fewer procedures might be needed related to fuel handling.

7.38. All personnel using operating procedures are required to be adequately trained in their use (see para. 7.61 of SSR-3 [1]). A graded approach could be used in application of this requirement. For example, in a research reactor with a high potential hazard, with complex SSCs important to safety, training on a specific procedure could involve extensive training on related SSCs. In comparison, training for the use of a simple procedure for the maintenance of a component in a research reactor with a low potential hazard could take less time.

7.39. Operating procedures need to be prepared, reviewed and submitted for approval on the basis of criteria established by the operating organization and regulatory requirements. This applies to all research reactors; however, the detail of the operating procedures can differ on the basis of their importance to safety, for example, as follows:

(a) The procedure for regeneration of an ion exchange system for producing demineralized water for a storage tank will be of low safety significance and will involve mature and simple technology. Consequently, the operating procedure governing this application can be simplified. In some cases, the ion exchange resins can be dried but radionuclides might be released during the drying process. As there are limits to be observed for radioactive discharges into the air, the safety significance is not to be graded as low in such cases.

(b) An operating procedure in which an error could cause a violation of the operational limits and conditions should be more detailed. An example is the procedure for regeneration of an ion exchange system for a primary cooling water purification system. While it involves the same basic technology as the example in point (a) above, the safety implications of an error could be much more significant (e.g. if resin were allowed to enter the primary cooling water and, hence, the reactor core). Design features and/or procedural arrangements should, therefore, take into account the greater hazard associated with the failure of this system, and the development, review and approval of operating procedures governing such safety significant activities should follow a stringent process.

(c) Procedures for changes in reactor utilization, special fuel tests, experiments and other special applications are often complex and infrequently used. Since these activities will often impact safety, the development, review and approval of procedures for these activities should follow the same process as that for other procedures governing safety significant activities.

Main control room, supplementary control room and control equipment

7.40. Requirements for the main control room, the supplementary control room and control equipment for research reactors are established in Requirement 75 of SSR-3 [1].

7.41. Paragraph 7.63 of SSR-3 [1] states:

"The habitability and good condition of control rooms shall be maintained. Where the design of the research reactor foresees additional or local control rooms that are dedicated to the control of experiments that could affect the reactor conditions, clear communication lines shall be developed for ensuring an adequate transfer of information to the operators in the main control room."

In a research reactor with a high potential hazard, the supplementary control room should include more monitoring and control equipment than a shutdown panel for a research reactor with a low or medium potential hazard. The number of shutdown panels in locations other than the control rooms should be commensurate with the potential hazard of the research reactor. The frequency and scope of tests performed by the operating organization to confirm that the supplementary control room and the shutdown panels are in a proper state of operational readiness should be commensurate with the nature of the equipment and the potential hazard of the research reactor.

Material conditions and housekeeping

7.42. Requirements for material conditions and housekeeping for research reactors are established in Requirement 76 of SSR-3 [1]. High standards of material conditions and housekeeping, including cleanliness, accessibility, adequate lighting, appropriate storage conditions, and identification and labelling of safety equipment, are required regardless of the potential hazard of the research reactor.

Maintenance, periodic testing and inspection

7.43. Requirements for maintenance, periodic testing and inspection for research reactors are established in Requirement 77 of SSR-3 [1]. Recommendations on maintenance, periodic testing and inspection for research reactors are provided in SSG-81 [3].

7.44. In accordance with Requirement 77 of SSR-3 [1], effective programmes for maintenance, periodic testing and inspection are required for all research reactors regardless of their potential hazards. The scope and extent of the programme, and the resources needed for planning, implementation and assessing the programme should be commensurate with the potential hazards of the research reactor and could vary significantly depending on the design, size and complexity of the reactor. For a research reactor with a low potential hazard and fewer SSCs important to safety, these activities can be performed by qualified operating personnel. In contrast, a dedicated maintenance group is typically needed for a large research reactor with more SSCs and a high potential hazard. The number of maintenance staff should also be commensurate with the potential hazard of the research reactor.

7.45. Three aspects of Requirement 77 should be applied using a graded approach: the development of procedures; the work permit system used to implement these procedures (see para. 7.69 of SSR-3 [1]); and the frequency of maintenance, periodic testing and inspection activities (see para. 7.72 of SSR-3 [1]). The graded approach should be based on the potential hazard of the research reactor, the safety significance of the SSCs involved, the complexity of the maintenance, periodic testing or inspection activity and the potential radiation risk associated with each activity.

7.46. In developing the procedures for maintenance, periodic testing and inspection, consideration should also be given to the experience of the staff and their familiarity with the SSCs. Recommendations on the application of a graded approach to the requirements for procedures are provided in paras 7.35–7.39.

7.47. When maintenance, periodic testing or inspection of an SSC is straightforward or operating experience indicates a high reliability of the SSC, a review of the frequency and details of the maintenance, periodic testing or inspection activity leading to a change in the procedure might be justified. However, a change in the procedure should be subject to the established process for preparation, review and approval.

7.48. The frequency of maintenance, periodic testing and inspection of individual SSCs is required to be adjusted on the basis of experience so as to ensure adequate reliability of SSCs important to safety (see para. 7.72 of SSR-3 [1]). For example, some instrumentation in the scram safety actuation system could need daily testing to demonstrate its functional operability and availability, whereas a sump pump could be tested at a lower frequency based on the results of the safety analysis.

7.49. A balance should be sought between the improvement in the detection of faults that is gained from more frequent testing against the risk that testing could be performed incorrectly and leave the SSC in a degraded state. This consideration also applies for periodic maintenance. The frequency of periodic maintenance might also depend on potential hazards, for example, the replacement frequency of SSCs subject to ageing degradation due to the level of radiation hazards.

7.50. The period for which an SSC is permitted to be out of service while reactor operation continues is usually stated in the operational limits and conditions for the research reactor and can be based on the availability requirement for the SSC from the safety analysis. Additional recommendations are provided in SSG-81 [3].

7.51. The use of a work permit system is required in all research reactors of all levels of potential hazard (see para. 7.69 of SSR-3 [1]). All work permits for activities with potential radiation risk should be reviewed by radiation protection personnel to ensure that doses from the activity are within prescribed limits and are as low as reasonably achievable. Further recommendations on radiation protection in research reactor operation are provided in SSG-85 [7].

7.52. Some maintenance, periodic testing and inspection activities are highly specialized and involve complex and sophisticated techniques; these activities are more likely to be necessary in more complex research reactor designs. Such activities are often performed by contracted experts external to the operating organization for the research reactor. Such outsourcing should be carefully considered by the operating organization to ensure that external support is secured and that resources will be available throughout the operating lifetime of the research reactor. Recommendations on the use of external contractors for the performance of maintenance, periodic testing and inspection are provided in SSG-81 [3].

Core management and fuel handling

7.53. Requirement 78 of SSR-3 [1] states that "**Core management and fuel handling procedures for a research reactor facility shall be established to ensure compliance with operational limits and conditions and consistency with the utilization programme.**" This requirement is applicable to all research reactors regardless of their potential hazard. In addition, the requirements on monitoring the integrity of the reactor core and the fuel and on the confinement of failed fuel (see para. 7.82 of SSR-3 [1]) apply equally to research reactors regardless of the potential hazard.

7.54. Recommendations on core management and fuel handling are provided in SSG-82 [4]. Research reactors with a low potential hazard and which involve infrequent changes to the core configuration might need a less comprehensive core management and fuel handling programme. Such research reactors operate with substantial margins to thermal limits, allowing the consideration of a broad envelope of acceptable fuel loading patterns in the initial safety analysis in lieu of core specific calculations. While all recommendations in SSG-82 [4] should be considered, some might not apply to these research reactors with a low potential hazard. Some research reactors, including some critical assemblies and subcritical assemblies, might undergo frequent changes to core configuration and fuel handling operations. As a result, these facilities will need a more comprehensive core management and fuel handling programme.

7.55. The safety significance of changes to research reactor core management and fuel handling procedures should be determined. SSG-24 (Rev. 1) [11] provides recommendations on a method for determining the safety significance of modifications to a research reactor and this method is applicable to core management and fuel handling. A graded approach to the analysis and verification of proposed changes to core management and fuel handling activities may be appropriate, on the basis of the safety significance of these changes (see also paras 7.70–7.74 of this Safety Guide).

7.56. A graded approach can also be used in determining the appropriate level of detail of the documentation and records on the status of fuel and core components (see para. 7.84 of SSR-3 [1]). Research reactors of a high potential hazard may need a more comprehensive system to document the status, and the evolution of this status with time, of each fuel assembly and core component, including experimental devices. In some research reactors with a high potential hazard, with complex systems and a diverse utilization programme, a dedicated group for core management and fuel handling may be necessary.

Fire safety

7.57. The requirements for fire safety for research reactors are established in Requirement 79 of SSR-3 [1]. Recommendations on fire safety are provided in IAEA Safety Standards Series No. SSG-77, Protection against Internal and External Hazards in the Operation of Nuclear Power Plants [35] and IAEA Safety Standards Series No. SSG-64, Protection Against Internal Hazards in the Design of Nuclear Power Plants [36]. Compliance with national requirements for fire safety cannot be subject to a graded approach.

7.58. The potential fire hazards should be discussed in the safety analysis report and an indication should be provided of their relative importance (i.e. in terms of likelihood and consequences) in the research reactor. This information can serve as a basis for the use of a graded approach in the implementation of the fire protection measures. For example, a fire affecting the instrumentation in the control room of a research reactor with a high potential hazard could be identified in the safety analysis as an event with a potential high consequence, needing to be mitigated by special means. A fire in an administrative area, with a low safety consequence identified in the safety analysis, could be mitigated by the deployment of hand-held fire extinguishers and the actions of firefighting personnel.

7.59. The use of a graded approach to implement the measures for fire protection might be facilitated by provisions incorporated into the design in accordance with the fire hazard analysis, which is required for all research reactors regardless of potential hazard (see para. 7.87 of SSR-3 [1]), and which should be periodically reviewed and updated (see SSG-77 [35]). The use of a graded approach to fire protection might also depend on the site of the research reactor.

7.60. Techniques for fire safety assessment and analysis are well understood; consequently, the amount of analysis needed on how best to apply the available resources can be determined using a graded approach. The analysis should employ techniques that have proven adequate in similar facilities elsewhere.

7.61. Recommendations on the use of a graded approach in the design of fire protection systems for research reactors are provided in para. 6.134.

Non-radiation-related safety

7.62. Requirements for a non-radiation-related safety programme for a research reactor are established in Requirement 80 of SSR-3 [1]. Each non-radiation-related hazard should be adequately addressed based on the nature of the hazard. The scope and level of detail of the programme should be developed using a graded approach based on the size and complexity of the research reactor and the specific hazards arising from its SSCs and operation.

Emergency preparedness

7.63. The requirements for emergency preparedness for research reactors are established in Requirement 81 of SSR-3 [1]. General requirements for emergency preparedness and response are established in IAEA Safety Standards Series No. GSR Part 7, Preparedness and Response for a Nuclear or Radiological Emergency [37].

7.64. Paragraph 7.90 of SSR-3 states:

"Emergency plans and procedures shall be based on the accidents analysed in the safety analysis report as well as those additionally postulated for the purposes of emergency preparedness and response on the basis of the hazard assessment."

The safety analysis and hazard assessment will allow the development of a source term for use in emergency planning. For some research reactors, it may be possible to demonstrate that the effects on the population and on the environment for credible accident scenarios are negligible and that emergency preparedness may be focused on the on-site response. An understanding of the nature and magnitude of the potential hazard posed by each research reactor, documented in a hazard assessment, is necessary for preparing an appropriate emergency plan and applying the requirements for emergency preparedness and response using a graded approach.

7.65. With regard to the application of a graded approach, para. 4.19 and table 1 of GSR Part 7 [37] establish a categorization scheme for facilities and activities to provide a basis for developing justified and optimized arrangements for emergency preparedness and response that are commensurate with the hazards. Most research reactor facilities are in emergency preparedness category II or III, depending on whether the research reactor can generate events that require an off-site response as well as an on-site response.

7.66. The magnitude of the potential source term, the proximity of the research reactor to population groups, and the engineered safety features are the most important factors to be considered in applying a graded approach to emergency planning, for example, in the following areas:

(a) The organizational structure needed to implement the emergency response.
(b) The size of the emergency planning zones.
(c) The identification of hazard and emergency classification.
(d) Notification of and communication with local, regional and national authorities, as appropriate.
(e) The amount, nature and storage location of equipment needed to survey and monitor people and the environment in the event of an emergency.
(f) The number and type of emergency services (e.g. police, firefighting services, ambulance services and medical facilities) the emergency response training of personnel in these organizations, and the nature of agreements with the operating organization. Even if an emergency might not have an off-site impact, it is prudent to establish contact with appropriate local, regional or national authorities at the planning stage to ensure their agreement if a request for assistance is issued.
(g) The timescales envisaged for each phase of the response to an emergency.
(h) The types, frequency and extent of training, exercises and drills in relation to on-site emergency response and (where needed) off-site emergency response.
(i) Any other resources needed for preparedness for and response to an emergency at the research reactor.

7.67. For a research reactor with a high potential hazard, there could be a need for a large amount of portable radiation monitoring equipment and emergency response equipment to be available at on-site locations. This equipment could also be used in emergency preparedness drills and in the training of on-site personnel and personnel from off-site organizations. For a research reactor with a lower potential hazard and no potential for off-site radiological consequences, far fewer portable radiation protection instruments and much less emergency equipment could be necessary for emergency response. In all cases, the equipment for use in emergency response is required to be maintained in good operational condition (see para. 7.93 of SSR-3 [1]) and should be included in the maintenance and periodic testing and inspection programme for the research reactor.

Records and reports

7.68. Requirements for records and reports for research reactors are established in Requirement 82 of SSR-3 [1]. Requirements for the control of records and documentation are also established in Requirements 8 and 10 of GSR Part 2 [14], and recommendations are provided in paras 5.35–5.49 of GS-G-3.1 [19]. The requirements for design information to be kept up to date for the duration of the operational stage of the research reactor and for information in logbooks and other records to be properly dated and signed (see paras 7.94 and 7.95 of SSR-3 [1], respectively), are applied irrespective of potential hazard of the facility.

7.69. Paragraph 2.44 of GS-G-3.1 [19] lists specific examples of where a graded approach could be applied to controls for records management, as follows:

(a)   Preparation of documents and records;
(b)   The need for and extent of validation of records and reports;
(c)   The degree of review and the individuals involved;
(d)   The level of approval for report and records;
(e)   The need for distribution lists;
(f)   The types of document that can be supplemented by temporary documents;
(g)   The need to archive superseded documents;
(h)   The need to categorize, register, index, retrieve and store records and reports;
(i)   The retention time for records;
(j)   Responsibilities for the disposal of records;
(k)   The types of storage medium.

Utilization and modification of a research reactor

7.70. Requirements for the utilization and modification of research reactors are established in Requirement 83 of SSR-3 [1]. Recommendations on the utilization and modification of research reactors are provided in SSG-24 (Rev. 1) [11].

7.71. The operating organization is required to establish criteria for categorizing a proposed experiment or modification in accordance with its importance to safety (see para. 7.100 of SSR-3 [1]). The resulting categorization should then be used to determine the types and extent of the analysis and approvals to be applied to the proposal. Annex I to SSG-24 (Rev. 1) [11] provides an example of a checklist for categorizing modifications in accordance with their potential hazard, using a safety screening checklist that divides modifications into four categories, as follows:

(a) Modifications with a major effect on safety;
(b) Modifications with a significant effect on safety;
(c) Modifications with a minor effect on safety;
(d) Modifications with no effect on safety.

7.72. Alternatively, a two-category system can be used for modifications to a research reactor. The first category is for modifications and experiments that are submitted to the regulatory body for review and approval. It includes modifications or experiments that (i) involve changes in the approved operational limits and conditions; (ii) affect items of major importance to safety; or (iii) involve hazards different in nature or more likely to occur than those previously considered. The second category is for modifications and experiments that need local review and approval, with notification to the regulatory body for information.

7.73. In cases where an experiment or modification was not anticipated and analysed in the design, its safety significance is required to be determined (see para. 7.99(a) of SSR-3 [1]).

7.74. The level of detail and depth of analysis that are necessary for the design, safety analysis, quality assurance, installation procedures, commissioning plan and training for personnel who will implement the modification as well as those who will use the SSC after modification can be implemented using a graded approach. Similarly, the scope and level of detail of the review performed by the regulatory body can be determined using a graded approach based on the safety significance of the modification.

Radiation protection programme

7.75. The requirements for a radiation protection programme at a research reactor are established in Requirement 84 of SSR-3 [1]. General requirements for radiation protection are established in IAEA Safety Standards Series No. GSR Part 3, Radiation Protection and Safety of Radiation Sources: International Basic Safety

Standards [38]. Recommendations on radiation protection in the design and operation of research reactors are provided in SSG-85 [7].

7.76. The application of the requirements for the radiation protection programme should be consistent with the reactor's design and its utilization. While the content of the radiation protection programme will depend on the design, power level, radiological hazards and utilization of the particular research reactor, many aspects of the programme should be similar for all research reactors. Paragraph 7.110 of SSR-3 [1] lists measures that are required in radiation protection programmes for research reactors with all levels of potential hazard.

7.77. The scope of environmental monitoring as part of the radiation protection programme (see para. 7.110(b) of SSR-3 [1]) is dependent on the potential hazard of the research reactor. For example, a research reactor located close to a densely populated area should be expected to undertake more extensive environmental monitoring.

7.78. Working areas within a research reactor are required to be designated as supervised areas or controlled areas (see para. 6.97 of SSR-3 [1] and paras 3.88–3.92 of GSR Part 3 [38]), in accordance with the magnitude of the expected exposures, the likelihood and magnitude of potential exposures, and the nature and extent of the radiological protection measures (see SSG-85 [7]).

7.79. For a research reactor with a high potential hazard, it may be necessary to further categorize the controlled areas into different levels, for example, levels I, II and III. Specific procedures may be prescribed for work in level II controlled areas (in addition to those procedures prescribed for level I controlled areas), which may involve, in some cases, the use of protective clothing, equipment or tools. Level III controlled areas should normally be accessed via a physical barrier (e.g. an airlock door) that can be opened only by authorized personnel. Furthermore, opening a door to a level III controlled area during reactor operation could be designed to result in automatic shutdown of the reactor. For a research reactor with a lower potential hazard, and a smaller number of areas where radiation hazards are present, the controlled area could be categorized into a smaller number of levels where additional radiation protection measures are needed. For a research reactor with a low potential hazard, with no locations where high dose rates are present, level II controlled areas and level III controlled areas may not be needed.

7.80. A critical assembly could present a higher risk of external radiation exposure of operating personnel than a higher power research reactor; however, the latter could present a higher risk of internal radiation exposure of operating personnel due to contamination. Critical assemblies are sometimes located within conventional industrial buildings; consequently, reactivity accidents involving a critical assembly could result in a higher risk of contamination outside the building compared with higher power research reactors with a larger source term that have a containment structure. These factors should be considered in the use of a graded approach in the application of the requirements for a radiation protection programme for a research reactor.

7.81. Allocating sufficient resources for the radiation protection programme to advise on and enforce radiation protection regulations, standards and procedures (see para. 7.108 of SSR-3 [1]), is an aspect of this requirement that can be applied using a graded approach. For example, at a research reactor with a high potential hazard and many SSCs with potential radiation hazards, the radiation protection group in the operating organization could include a large number of personnel, working in shifts, trained to use a number of instruments for detecting and characterizing sources of radiation, and involved in the planning and execution of activities in the research reactor. In a research reactor with a low potential hazard, such as some critical assemblies and subcritical assemblies, radiation protection tasks could be performed by just one or two personnel who are also trained in other operational activities.

Management of radioactive waste

7.82. General requirements for the predisposal management of radioactive waste are established in GSR Part 5 [32]. Requirements for the management of radioactive waste generated at research reactors are established in Requirement 85 of SSR-3 [1]. Recommendations on the management of such radioactive waste at research reactors are provided in SSG-85 [7]. The safety of radioactive waste management activities should be subject to safety assessment and periodic safety reviews (see Requirements 13–16 of GSR Part 5 [32]). The operating organization should use safety assessments to inform the design of waste management activities so that they are appropriate to the hazard posed by the waste in question.

7.83. The operating organization should establish a radioactive waste management programme for the research reactor site and implement this programme in accordance with a graded approach. The scope of the radioactive waste management programme should be consistent with the size and complexity of the research reactor operations. Requirement 85 of SSR-3 [1] can be applied using a graded approach based on the quantity and characteristics of radioactive waste generated and the associated licence conditions. For a research reactor with a high potential hazard, there might be a diverse range of radioactive waste generated, including waste oil from maintenance activities, liquid and gaseous effluents from reactor operation, solid and liquid waste from isotope production, and contaminated disposable materials from radiation protection and decontamination activities. In contrast, the quantity of waste generated, and the associated radiation risk for a research reactor with a low potential hazard are typically less.

7.84. Paragraph 7.116 of SSR-3 [1] states:

"The reactor and its experimental devices shall be operated to minimize the generation of radioactive waste of all kinds, to ensure that releases of radioactive material to the environment are kept below permissible regulatory limits and as low as reasonably achievable and to facilitate the handling and disposal of waste."

This requirement applies equally to all research reactors.

Ageing management

7.85. Requirements for an ageing management programme for research reactors are established in Requirement 86 of SSR-3 [1]. Recommendations on ageing management for research reactors are provided in SSG-10 (Rev. 1) [8].

7.86. Aspects of the requirement for an ageing management programme for a research reactor that can be applied using a graded approach include the following:

(a) The frequency of inspections for the detection of ageing effects;
(b) The resources necessary to implement an ageing management programme;
(c) The implementation of corrective actions resulting from a periodic safety review (see para. 7.121 of SSR-3 [1]).

7.87. The appropriate frequency of inspections, and the measures for mitigation of ageing effects, should be based on the importance to safety, estimated service life, complexity and ease of replacement of individual SSCs. In most research reactors, it is feasible to inspect most SSCs periodically and to replace components, if necessary. Inspections should be prioritized where degradation mechanisms have been identified. The SSCs that perform the main safety functions should be prioritized for ageing management inspections.

7.88. The allocation of the resources necessary to implement an ageing management programme for a research reactor can also be based on a graded approach. For a research reactor with a high potential hazard, a dedicated organizational unit might be needed to implement such a programme, including planning and performing ageing management activities in coordination with the maintenance programme (see para. 7.120 of SSR- 3 [1]). For a research reactor with a low potential hazard, the ageing management programme activities might be planned, supervised and performed by the maintenance personnel in the operating organization.

7.89. The implementation of corrective actions resulting from a periodic safety review can be applied using a graded approach. The assessment of the findings from the review should apply risk based significance levels to all proposed corrective actions. The operating organization might decide not to implement a corrective action for an issue of low safety significance where there is sufficient justification. This approach to the corrective actions from a periodic safety review is applicable to all research reactors regardless of their potential hazard.

Extended shutdown

7.90. Requirements for the extended shutdown of a research reactor are established in Requirement 87 of SSR-3 [1]. The recommendations on long shutdown provided in paras 6.74 and 6.75 of this Safety Guide are also applicable during the extended shutdown of a research reactor.

7.91. Requirement 2 of SSR-3 [1] states that "**The operating organization for a research reactor facility shall have the prime responsibility for the safety of the research reactor over its lifetime**". This responsibility remains during the period of extended shutdown (i.e. while the decision has not been made to either decommission or restart the research reactor).

7.92. A graded approach should be applied to the activities, the measures to be implemented, the level of reviews, the frequency and extent of maintenance, and the testing and inspection activities during an extended shutdown, and the extent of relief from requirements that apply during the normal operating regime, including any licence conditions and operational limits and conditions. Any such relief should be subjected to safety analysis and regulatory review and assessment.

Feedback of operating experience

7.93. Requirements for feedback of operating experience of a research reactor are established in Requirement 88 of SSR-3 [1]. The requirement for the operating organization to report, collect, screen, analyse, trend, document and communicate operating experience at the reactor facility in a systematic way (see para. 7.126 of SSR-3 [1]) applies regardless of the potential hazard of the research reactor.

7.94. The resources necessary to implement the operating experience programme, and the scope of this programme, should be commensurate with the potential hazard of the research reactor, the number and complexity of SSCs important to safety and the size of the operating organization.

8. USE OF A GRADED APPROACH IN THE PREPARATION FOR DECOMMISSIONING OF RESEARCH REACTORS

8.1. Requirement 89 of SSR-3 [1] states:

> **"The operating organization for a research reactor facility shall prepare a decommissioning plan and shall maintain it throughout the lifetime of the research reactor, unless otherwise approved by the regulatory body, to demonstrate that decommissioning can be accomplished safely and in such a way as to meet the specified end state."**

8.2. The scope, extent, and level of detail of the safety assessment for decommissioning and the decommissioning plan should be commensurate with the hazards associated with the decommissioning of the research reactor. The effort associated with meeting the requirements for the preparation and review of decommissioning plans and procedures should also be based on the potential hazards associated with the decommissioning of the facility. Depending on these hazards, and on the design, complexity and history of operation and utilization of the facility, a graded approach can be used to determine the most appropriate level and depth of analyses, the type and number of decommissioning procedures to be prepared as well as the scope and depth of safety reviews and assessments. A graded approach should also be used in determining the appropriate type, extent and level of detail of surveillance and radiation protection measures, including monitoring, during transition from operation to decommissioning.

8.3. Preparation for decommissioning should include a consideration of the knowledge of the research reactor that might be lost as a result of the loss of experienced personnel when the reactor is permanently shut down. The requirement for the operating organization to retain personnel and preserve knowledge of the research reactor (see para. 8.7 of SSR-3 [1]) should be applied using a graded approach, based on the potential hazards, the knowledge of the facility and its safety significance to decommissioning. For research reactors with a small operating organization, preserving the knowledge of a small number of key personnel may be essential for preparation for decommissioning.

8.4. A graded approach should be applied to the scope and level of details of the decommissioning plan, based on the potential hazard of the shutdown research reactor (e.g. with nuclear fuel removed), the resources available for decommissioning, the time period to decommissioning and the specified end state of the facility (e.g. full or partial decontamination and/or dismantling or release of the site from regulatory control). Requirements for the decommissioning of facilities, including research reactors, are established in IAEA Safety Standards Series No. GSR Part 6, Decommissioning of Facilities [39] and recommendations are provided in IAEA Safety Standards Series No. SSG-47, Decommissioning of Nuclear Power Plants, Research Reactors and Other Nuclear Fuel Cycle Facilities [40].

9. USE OF A GRADED APPROACH IN THE INTERFACES BETWEEN SAFETY AND SECURITY FOR RESEARCH REACTORS

9.1. Requirement 90 of SSR-3 [1] states:

"The interfaces between safety and security for a research reactor facility shall be addressed in an integrated manner throughout the lifetime of the reactor. Safety measures and security measures shall be established and implemented in such a manner that they do not compromise one another."

9.2. The requirement that safety and security issues are addressed in an integrated manner is applied irrespective of the potential hazard of the facility. Safety and security are two distinct areas essential for the operation of a research reactor. A graded approach can be used in the activities necessary for the effective management of the interface between safety and security. This includes the following:

(a) The number and extent of coordinated safety and security regulatory inspections and emergency drills;

(b)   The extent and level of detail of review of the access control procedures by safety specialists;

(c)   The extent and level of detail of review of the operating and maintenance procedures by security specialists;

(d)   The extent of reviews by security specialists of modifications important to safety;

(e)   The extent of reviews of modifications of nuclear security systems by safety specialists while ensuring appropriate information security;

(f)   The contents of training of safety aspects for security specialists and vice versa.

9.3.   Recommendations related to the interfaces between safety and security are included in the Safety Guides listed in para. 1.3, in particular SSG-84 [6] and SSG-24 (Rev. 1) [11].[6]

REFERENCES

[1]   INTERNATIONAL ATOMIC ENERGY AGENCY, Safety of Research Reactors, IAEA Safety Standards Series No. SSR-3, IAEA, Vienna (2016).

[2]   INTERNATIONAL ATOMIC ENERGY AGENCY, Commissioning of Research Reactors, IAEA Safety Standards Series No. SSG-80, IAEA, Vienna (in preparation).

[3]   INTERNATIONAL ATOMIC ENERGY AGENCY, Maintenance, Periodic Testing and Inspection of Research Reactors, IAEA Safety Standards Series No. SSG-81, IAEA, Vienna (in preparation).

[4]   INTERNATIONAL ATOMIC ENERGY AGENCY, Core Management and Fuel Handling for Research Reactors, IAEA Safety Standards Series No. SSG-82, IAEA, Vienna (in preparation).

[5]   INTERNATIONAL ATOMIC ENERGY AGENCY, Operational Limits and Conditions and Operating Procedures for Research Reactors, IAEA Safety Standards Series No. SSG-83, IAEA, Vienna (in preparation).

[6]   INTERNATIONAL ATOMIC ENERGY AGENCY, The Operating Organization and the Recruitment, Training and Qualification of Personnel for Research Reactors, IAEA Safety Standards Series No. SSG-84, IAEA, Vienna (in preparation).

[7]   INTERNATIONAL ATOMIC ENERGY AGENCY, Radiation Protection and Radioactive Waste Management in the Design and Operation of Research Reactors, IAEA Safety Standards Series No. SSG-85, Vienna (in preparation).

---

[6]   Practical guidance on the use of a graded approach and the safety and security interface is provided in Ref. [41].

[8]   INTERNATIONAL ATOMIC ENERGY AGENCY, Ageing Management for Research Reactors, IAEA Safety Standards Series No. SSG-10 (Rev. 1), IAEA, Vienna (in preparation).

[9]   INTERNATIONAL ATOMIC ENERGY AGENCY, Instrumentation and Control Systems and Software Important to Safety for Research Reactors, IAEA Safety Standards Series No. SSG-37 (Rev. 1), IAEA, Vienna (in preparation).

[10]   INTERNATIONAL ATOMIC ENERGY AGENCY, Safety Assessment for Research Reactors and Preparation of the Safety Analysis Report, IAEA Safety Standards Series No. SSG-20 (Rev. 1), IAEA, Vienna (2022).

[11]   INTERNATIONAL ATOMIC ENERGY AGENCY, Safety in the Utilization and Modification of Research Reactors, IAEA Safety Standards Series No. SSG-24 (Rev. 1), IAEA, Vienna (2022).

[12]   INTERNATIONAL ATOMIC ENERGY AGENCY, IAEA Nuclear Safety and Security Glossary: Terminology Used in Nuclear Safety, Nuclear Security, Radiation Protection and Emergency Preparedness and Response, 2022 (Interim) Edition, IAEA, Vienna (2022).

[13]   INTERNATIONAL ATOMIC ENERGY AGENCY, Establishing the Safety Infrastructure for a Nuclear Power Programme, IAEA Safety Standards Series No. SSG-16 (Rev. 1), IAEA, Vienna (2020).

[14]   INTERNATIONAL ATOMIC ENERGY AGENCY, Leadership and Management for Safety, IAEA Safety Standards Series No. GSR Part 2, IAEA, Vienna (2016).

[15]   INTERNATIONAL ATOMIC ENERGY AGENCY, Safety Assessment for Facilities and Activities, IAEA Safety Standards Series No. GSR Part 4 (Rev. 1), IAEA, Vienna (2016).

[16]   INTERNATIONAL LABOUR ORGANIZATION, Guidelines on Occupational Safety and Health Management Systems, ILO-OSH 2001, ILO, Geneva (2001).

[17]   INTERNATIONAL LABOUR ORGANIZATION, Safety and Health in Construction, An ILO Code of Practice, ILO, Geneva (1992).

[18]   INTERNATIONAL LABOUR ORGANIZATION, Safety in the Use of Chemicals at Work, An ILO Code of Practice, ILO, Geneva (1993).

[19]   INTERNATIONAL ATOMIC ENERGY AGENCY, Application of the Management System for Facilities and Activities, IAEA Safety Standards Series No. GS-G-3.1, IAEA, Vienna (2006). (A revision of this publication is in preparation.)

[20]   INTERNATIONAL ATOMIC ENERGY AGENCY, Governmental, Legal and Regulatory Framework for Safety, IAEA Safety Standards Series No. GSR Part 1 (Rev. 1), IAEA, Vienna (2016).

[21]   INTERNATIONAL ATOMIC ENERGY AGENCY, Functions and Processes of the Regulatory Body for Safety, IAEA Safety Standards Series No. GSG-13, IAEA, Vienna (2018).

[22]   INTERNATIONAL ATOMIC ENERGY AGENCY, Application of a Graded Approach in Regulating Nuclear Installations, IAEA-TECDOC-1980, IAEA, Vienna (2021).

[23]   INTERNATIONAL ATOMIC ENERGY AGENCY, Organization, Management and Staffing of the Regulatory Body for Safety, IAEA Safety Standards Series No. GSG-12, IAEA, Vienna (2018).

[24] INTERNATIONAL ATOMIC ENERGY AGENCY, Site Evaluation for Nuclear Installations, IAEA Safety Standards Series No. SSR-1, IAEA, Vienna (2019).

[25] INTERNATIONAL ATOMIC ENERGY AGENCY, Site Survey and Site Selection for Nuclear Installations, IAEA Safety Standards Series No. SSG-35, IAEA, Vienna (2015).

[26] INTERNATIONAL ATOMIC ENERGY AGENCY, Seismic Hazards in Site Evaluation for Nuclear Installations, IAEA Safety Standards Series No. SSG-9 (Rev. 1), IAEA, Vienna (2022).

[27] INTERNATIONAL ATOMIC ENERGY AGENCY, Volcanic Hazards in Site Evaluation for Nuclear Installations, IAEA Safety Standards Series No. SSG-21, IAEA, Vienna (2012).

[28] INTERNATIONAL ATOMIC ENERGY AGENCY, WORLD METEOROLOGICAL ORGANIZATION, Meteorological and Hydrological Hazards in Site Evaluation for Nuclear Installations, IAEA Safety Standards Series No. SSG-18, IAEA, Vienna (2011).

[29] INTERNATIONAL ATOMIC ENERGY AGENCY, Hazards Associated with Human Induced External Events in Site Evaluation for Nuclear Installations, IAEA Safety Standards Series No. SSG-79, IAEA, Vienna (2023).

[30] INTERNATIONAL ATOMIC ENERGY AGENCY, The Management System for Nuclear Installations, IAEA Safety Standards Series No. GS-G-3.5, IAEA, Vienna (2009).

[31] INTERNATIONAL ATOMIC ENERGY AGENCY, Safety Analysis for Research Reactors, Safety Report Series No. 55, IAEA, Vienna (2008).

[32] INTERNATIONAL ATOMIC ENERGY AGENCY, Predisposal Management of Radioactive Waste, IAEA Safety Standards Series No. GSR Part 5, IAEA, Vienna (2009).

[33] INTERNATIONAL ATOMIC ENERGY AGENCY, Regulations for the Safe Transport of Radioactive Material, 2018 Edition, IAEA Safety Standards Series No. SSR-6 (Rev. 1), IAEA, Vienna (2018).

[34] INTERNATIONAL ATOMIC ENERGY AGENCY, The Management System for the Safe Transport of Radioactive Material, IAEA Safety Standards Series No. TS-G-1.4, IAEA, Vienna (2008). (A revision of this publication is in preparation.)

[35] INTERNATIONAL ATOMIC ENERGY AGENCY, Protection Against Internal and External Hazards in the Operation of Nuclear Power Plants, IAEA Safety Standards Series No. SSG-77, IAEA, Vienna (2022).

[36] INTERNATIONAL ATOMIC ENERGY AGENCY, Protection Against Internal Hazards in the Design of Nuclear Power Plants, IAEA Safety Standards Series No. SSG-64, IAEA, Vienna (2021).

[37] FOOD AND AGRICULTURE ORGANIZATION OF THE UNITED NATIONS, INTERNATIONAL ATOMIC ENERGY AGENCY, INTERNATIONAL CIVIL AVIATION ORGANIZATION, INTERNATIONAL LABOUR ORGANIZATION, INTERNATIONAL MARITIME ORGANIZATION, INTERPOL, OECD NUCLEAR ENERGY AGENCY, PAN AMERICAN HEALTH ORGANIZATION, PREPARATORY COMMISSION FOR THE COMPREHENSIVE NUCLEAR-TEST-BAN TREATY ORGANIZATION, UNITED NATIONS ENVIRONMENT PROGRAMME, UNITED NATIONS OFFICE FOR THE COORDINATION OF HUMANITARIAN AFFAIRS, WORLD HEALTH ORGANIZATION, WORLD METEOROLOGICAL ORGANIZATION, Preparedness and Response for a Nuclear or Radiological Emergency, IAEA Safety Standards Series No. GSR Part 7, IAEA, Vienna (2015).

[38] EUROPEAN COMMISSION, FOOD AND AGRICULTURE ORGANIZATION OF THE UNITED NATIONS, INTERNATIONAL ATOMIC ENERGY AGENCY, INTERNATIONAL LABOUR ORGANIZATION, OECD NUCLEAR ENERGY AGENCY, PAN AMERICAN HEALTH ORGANIZATION, UNITED NATIONS ENVIRONMENT PROGRAMME, WORLD HEALTH ORGANIZATION, Radiation Protection and Safety of Radiation Sources: International Basic Safety Standards, IAEA Safety Standards Series No. GSR Part 3, IAEA, Vienna (2014).

[39] INTERNATIONAL ATOMIC ENERGY AGENCY, Decommissioning of Facilities, IAEA Safety Standards Series No. GSR Part 6, IAEA, Vienna (2014).

[40] INTERNATIONAL ATOMIC ENERGY AGENCY, Decommissioning of Nuclear Power Plants, Research Reactors and Other Nuclear Fuel Cycle Facilities, IAEA Safety Standards Series No. SSG-47, IAEA, Vienna (2018).

[41] INTERNATIONAL ATOMIC ENERGY AGENCY, Management of the Interface Between Nuclear Safety and Security for Research Reactors, IAEA-TECDOC-1801, IAEA, Vienna (2016).

CONTRIBUTORS TO DRAFTING AND REVIEW

Abou Yehia, H.	Consultant, France
Barnea, Y	Consultant, Israel
D'Arcy, A	Consultant, South Africa
Helvenston, E.	Nuclear Regulatory Commission, United States of America
McIvor, A.	International Atomic Energy Agency
Naseer, F.	International Atomic Energy Agency
Sears, D.	International Atomic Energy Agency
Shokr, A.M.	International Atomic Energy Agency
Waldman, R.	Consultant, Argentina

ORDERING LOCALLY

IAEA priced publications may be purchased from the sources listed below or from major local booksellers.

Orders for unpriced publications should be made directly to the IAEA. The contact details are given at the end of this list.

NORTH AMERICA

Bernan / Rowman & Littlefield
15250 NBN Way, Blue Ridge Summit, PA 17214, USA
Telephone: +1 800 462 6420 • Fax: +1 800 338 4550
Email: orders@rowman.com • Web site: www.rowman.com/bernan

REST OF WORLD

Please contact your preferred local supplier, or our lead distributor:

Eurospan Group
Gray's Inn House
127 Clerkenwell Road
London EC1R 5DB
United Kingdom

Trade orders and enquiries:
Telephone: +44 (0)176 760 4972 • Fax: +44 (0)176 760 1640
Email: eurospan@turpin-distribution.com

Individual orders:
www.eurospanbookstore.com/iaea

For further information:
Telephone: +44 (0)207 240 0856 • Fax: +44 (0)207 379 0609
Email: info@eurospangroup.com • Web site: www.eurospangroup.com

Orders for both priced and unpriced publications may be addressed directly to:
Marketing and Sales Unit
International Atomic Energy Agency
Vienna International Centre, PO Box 100, 1400 Vienna, Austria
Telephone: +43 1 2600 22529 or 22530 • Fax: +43 1 26007 22529
Email: sales.publications@iaea.org • Web site: www.iaea.org/publications